Gems of Geometry

John Barnes

Gems of Geometry

Second Edition

John Barnes
Caversham, England

ISBN 978-3-642-30963-2   ISBN 978-3-642-30964-9 (eBook)
DOI 10.1007/978-3-642-30964-9
Springer Heidelberg New York Dordrecht London

Library of Congress Control Number: 2012946175

Mathematics Subject Classification: 00-XX, 37-XX, 51-XX, 54-XX, 83-XX

© Springer-Verlag Berlin Heidelberg 2012
This work is subject to copyright. All rights are reserved by the Publisher, whether the whole or part of the material is concerned, specifically the rights of translation, reprinting, reuse of illustrations, recitation, broadcasting, reproduction on microfilms or in any other physical way, and transmission or information storage and retrieval, electronic adaptation, computer software, or by similar or dissimilar methodology now known or hereafter developed. Exempted from this legal reservation are brief excerpts in connection with reviews or scholarly analysis or material supplied specifically for the purpose of being entered and executed on a computer system, for exclusive use by the purchaser of the work. Duplication of this publication or parts thereof is permitted only under the provisions of the Copyright Law of the Publisher's location, in its current version, and permission for use must always be obtained from Springer. Permissions for use may be obtained through RightsLink at the Copyright Clearance Center. Violations are liable to prosecution under the respective Copyright Law.
The use of general descriptive names, registered names, trademarks, service marks, etc. in this publication does not imply, even in the absence of a specific statement, that such names are exempt from the relevant protective laws and regulations and therefore free for general use.
While the advice and information in this book are believed to be true and accurate at the date of publication, neither the authors nor the editors nor the publisher can accept any legal responsibility for any errors or omissions that may be made. The publisher makes no warranty, express or implied, with respect to the material contained herein.

Printed on acid-free paper

Springer is part of Springer Science+Business Media (www.springer.com)

To
Bobby,
Janet, Helen,
Christopher and Jonathan

Preface

It is now three years since the first edition of this book was published in 2009. Moreover, I have recently given the lectures on which it is based twice more as part of the Continuing Education program of Oxford University in England. Experience of doing this gave me valuable feedback and suggested a little more material which could be added and so I was delighted when Springer asked me to prepare a second edition. Second editions are also an opportunity to remove blunders in the first edition (and hopefully not add too many more). Also, I am told that this new edition will be produced using print-on-demand technology and so can remain in print for evermore!

The first edition was based on lectures given several times at Reading University in England. One might wonder why I gave these lectures. I had been attending a few lectures on diverse subjects such as Music, Latin and Greek and my wife suggested that perhaps I should give some lectures myself. I was somewhat taken aback by this but then realised that I did have some useful material lying around that would make a starting point.

When I was a boy (some time ago) I much enjoyed reading books such as *Mathematical Snapshots* by Steinhaus and *Mathematical Recreations and Essays* by Rouse Ball. Moreover, I had made a few models such as the minimal set of squares that fit together to make a rectangle, some sets of Chinese Rings, and the MacMahon coloured cubes. This starting point was much enhanced by some models that my daughter Janet had made when at school. A junior class had been instructed to make some models for an open day but had made a mess instead. Janet (then in the sixth form) was asked to save the day. Thus I also had available models of many regular figures including the Poinsot-Kepler figures and the compound of five tetrahedra and that of five cubes.

Having written a few technical books on programming languages, it seemed a natural step to prepare colourful notes for the course in the form of what might be chapters of a book. It took longer than I thought to finally turn the notes into a book – partly because I was diverted into writing a number of other books on programming in the meantime. But at last the job was done. So that is the background to the first edition.

There are ten basic lectures. We start with the Golden Number which leads naturally to considering regular Shapes and Solids in two and three dimensions and then a foray into the Fourth Dimension. A little amusement with Projective Geometry follows (a necessity in my youth for those going to university) and then a dabble in Topology. A messy experience with soap Bubbles stimulated by Boys' little book is next. We then look at circles and spheres (the Harmony of the Spheres) and especially Steiner's porism and Soddy's hexlet, which provide opportunities for pretty diagrams. Next is a look at some aspects of Chaos and

Fractals. We then with some trepidation look at Relativity – special relativity can be appreciated relatively easily (groan – sorry about the pun) but general relativity is a bit tricky. The Finale then picks up a few loose ends.

Additional material on topics such as stereo images, further aspects of the fourth dimension, Schlegel diagrams and crystals has been gathered into a number of appendices. The main lectures contain some exercises (harder ones are marked with asterisks) but answers are not provided since I anticipate giving the course again.

The main changes in this second edition are that more material has been added on the golden number, shapes and solids, topology, chaos, and crystals. Also the colours of some diagrams, particularly those in stereo, have been changed to give a crisper effect. In particular, the appendix on crystals has been considerably extended to give a better treatment of (including better stereo images of) the structure of diamonds, graphite and a number of siliceous minerals such as quartz that exhibit intriguing geometrical properties.

An important question is to consider who might want to read this book. The mathematical background required is not hard (a bit of simple algebra, Pythagoras, a touch of trigonometry) and is the sort of stuff anyone who studied a science based subject to the age of 16 or so would have encountered. One obvious group therefore is young people with a zest for knowledge with beauty (I would have loved to have been given such a book when I was 16). Another group as evidenced by students on courses is those of maturer years who might like to know more about topics that they enjoyed when young.

I find that students on the courses are from varied backgrounds – of both sexes and all ages. Some have little technical background at all but revel in activities such as making models, cutting up Möbius strips and blowing bubbles; others have serious scientific experience and enjoy perhaps a nostalgic trip visiting some familiar topics and meeting fresh ones.

I have made no attempt to avoid using mathematical notation wherever it is appropriate. I have some objection to popular mathematical books that strive to avoid mathematics because some publisher once said that every time an equation is added, the sales divide by two. But I have aimed to provide lots of illustrations to enliven the text.

I must now thank all those who have helped me in this task. First, a big thank you to my wife, Bobby, who suggested giving the courses, helped with typesetting and took some of the photographs, and to my daughter Janet who provided much background material. Thanks also to David Shorter who took some other photographs; and to Frank Bott who translated parts of an ancient book in Italian; also to Brian Wichmann who gave good advice on generating some diagrams; and to Pascal Leroy who was a great help in finding a number of errors and suggesting many improvements.

I must especially pay tribute to the late John Dawes whom I knew when we were both undergraduates at Trinity in Cambridge – John worked with me in the software industry and provided the inspiration for some of the exercises. One episode is worth mentioning. We encountered the problem of proving that if one puts squares on the sides of any quadrilateral then the lines joining the centres of opposite squares are always the same length and at right angles. I challenged the team working on a large compiler project to find a slick proof. John came up with the proof using complex numbers described in the Finale.

I am grateful to the authors of the many books that I have read and enjoyed and which have been a real stimulus to understanding. I cannot mention them all here but I must mention a few. The oldest is probably *Flatland* by Edwin Abbott, a marvellous tale written over a century ago about adventures into many dimensions. And then from the same era there is *Soap Bubbles* by Boys with its elegant diagrams – it seems that he entertained Victorian dinner parties with his demonstrations. The various books by Coxeter such as *Introduction to Geometry* and *Regular Polytopes* are fascinating. Three other books I must mention. One is *Excursions in Geometry* by C Stanley Ogilvy which introduced me to Soddy's amazing hexlet; another is *Stamping Through Mathematics* by Robin Wilson which explores the world of mathematics via illustrations on postage stamps; the third is the *Emporer's New Mind* by Roger Penrose with its material on curious topics such as aperiodic tilings and pentagonal crystals.

In a nostalgic mood, I must thank those who taught me about the wonders of mathematics at school and at Cambridge. At school (Latymer Upper in West London), we were privileged to learn from brilliant teachers such as Bob Whittaker. I recall sixth form lessons in the basement of a local milkbar where we could contemplate projective geometry. At Cambridge, I especially enjoyed lectures by Fred Hoyle on relativity and Paul Dirac on quantum mechanics. I was also privileged to enjoy wonderful supervisions from John Polkinghorne who also very kindly reviewed part of the first edition of this book.

For the second edition, I must especially thank Pascal Leroy for a suggestion regarding dipoles and Roger Penrose for help in developing a brief description of aperiodic pentagonal tilings and suggesting a number of improvements to the discussion on semi-regular figures.

Finally, I must continue to thank my friends Karen Mosman, Sally Mortimore, and Simone Taylor who encouraged me to persevere with progressing the original book to publication and, most important of all, Martin Peters, Ruth Allewelt and Angela Schulze-Thomin at Springer-Verlag and Sorina Moosdorf at le-tex who make it actually happen.

I hope that all those who read or browse through this book will find something to enjoy. I enjoyed writing it and learnt a lot in the process.

John Barnes
Caversham
England
May 2012

Contents

1 The Golden Number   1
Pieces of paper, The golden ratio, Fibonacci's rabbits, Continued fractions, Pentagons, Phyllotaxis, Further reading, Exercises.

2 Shapes and Solids   27
Flatland, Polygons, Tiling, Vision and projection, Five classical polyhedra, Duality, Kepler and Poinsot, The Archimedean figures, Non-convex polyhedra, Pentagonal tilings, Further reading, Exercises.

3 The Fourth Dimension   63
What is the fourth dimension? Honeycombs, The 4-simplex, The hypercube and the 16-cell, Other regular convex figures, Non-convex regular figures, Honeycombs, five dimensions and more, Nets, Further reading, Exercises.

4 Projective Geometry   89
Pappus' theorem, Desargues' theorem, Duality, Duality in three dimensions, Infinity and parallels, Quadrilaterals and quadrangles, Conics, Coordinates, Finite geometries, Configurations, Further reading, Exercises.

5 Topology   113
Hairy dogs, Colour problems, Colouring maps on the torus, The Möbius band, The Klein bottle, The projective plane, Round up, Further reading, Exercises.

6 Bubbles   137
Surface tension, Two bubbles, Three bubbles, Four bubbles, Foam, Films on frames, Films on cylinders, That well known theorem, Further reading, Exercises.

7 Harmony of the Spheres — 157
Steiner's porism, Inversion, Coaxial circles, Proof of Steiner's porism, Soddy's hexlet, Further reading, Exercises.

8 Chaos and Fractals — 179
Shaken foundations, Fractals, Fractional dimensions, Cantor sets, Population growth, Double, double, boil and trouble, Chaos and peace, And so to dust, Newton's method, Julia and Mandelbrot sets, Natural chaos, Further reading, Exercises.

9 Relativity — 205
The special theory, Time changes, The Lorentz–Fitzgerald contraction, Distortion of bodies, Lorentz transformation, Time and relativity, Mass and energy, Coordinates, Curvature, Einstein's equations, The Schwarzchild solution, Consequences of general relativity, Black holes, Properties of black holes, Further reading, Exercise.

10 Finale — 231
Squares on a quadrilateral, The Argand plane, The quadrilateral revisited, Other complex problems, Trisection, Bends, Pedal triangles, Coordinates of points and lines, Further reading.

A The Bull and the Man — 247
The problem, The proof.

B Stereo Images — 251
Compound figures, Desargues' theorem.

C More on Four — 261
Archimedean figures in four dimensions, Prisms and hyperprisms.

D Schlegel Images — 271
Schlegel diagrams, The hypercube, The 16-cell, The 24-cell, The 120-cell, 600-cell and tetroctahedric.

E	**Crystals**	**287**

Packing of spheres, Crystals, Diamonds and graphite, Silica structures, Gems, Further reading.

F	**Stability**	**305**

Stability of fixed points, The fixed points, The two cycle, Three cycles.

G	**Fanoland**	**319**

Seven girls and seven boys, Plus six girls and six boys, Explanation.

Bibliography	321
Index	323

1 The Golden Number

THIS LECTURE is about the so-called Golden Number. This number defines a ratio which turns up in various guises in an amazing number of contexts. These include the ideal shapes of architectural objects such as temples, the arrangement of many botanical systems, and purely geometrical objects such as the pentagon. We shall start, however, by considering the sizes of pieces of paper.

Pieces of paper

ONCE UPON A TIME paper came in all sorts of bizarre sizes with wonderful names such as Elephant, Foolscap and Crown. Books were printed in sizes such as Crown octavo which had eight pages to a sheet of Crown. However, in the 1960s, as part of the metrication process, the A series of paper sizes was introduced and all these names faded (the A series was originally a German DIN standard). But note that in this respect the US is stuck in a time warp and continues to use historical sizes known as Letter and Legal.

So now we use paper with boring names such as A3 and A4. A sheet of A3 is simply double a sheet of A4 and similarly A5 is half a sheet of A4. Thus, as we add one to the number, the area is divided by two. The diagram shows a sheet of A3 divided into two sheets of A4.

An important property of the A series is that all the sizes have the same proportions. Suppose the width of a sheet of A4 is x and its length is y, then a sheet of A3 will have width y and length $2x$. If the ratios of the length to width for both sizes are the same, it follows that

$2x/y = y/x$, so that $(y/x)^2 = 2$ giving a ratio of $\sqrt{2} = 1.414...$

So each sheet has sides $\sqrt{2}$ smaller than the one before and area one-half the one before. Well this is well known. But what is less well known is that the area of A0, the biggest of the series, is defined to be exactly one square metre.

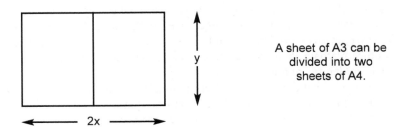

A sheet of A3 can be divided into two sheets of A4.

2 Gems of Geometry

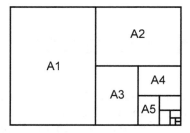

A sheet of A0 subdivided into an infinite sequence of smaller sheets.

This means that the sides of a sheet of A0 have to be $\sqrt{\sqrt{2}}$ and $1/\sqrt{\sqrt{2}}$ respectively which give a size of about 1189 by 841 mm. A4 has sides one-quarter of this giving 297 by 210 which is the right answer.

Note that a sheet of A0 can be subdivided into one each of the (infinite number of) smaller sizes as above.

This shows quite clearly that the sum of the areas of the smaller sheets equals the area of the big sheet of A0. And so it follows that

$$1/2 + 1/4 + 1/8 + 1/16 + 1/32 + \ldots = 1$$

This is therefore a very simple illustration of the fact that the sum of an infinite series of numbers can be finite.

This, of course, is the essence of Zeno's paradox about Achilles and the tortoise. Remember that they have a race and Achilles gives the tortoise 100 metres start. Achilles runs 10 times as fast as the tortoise so by the time Achilles reaches the point from whence the tortoise started, the tortoise is now 10 metres ahead; when Achilles reaches that point the tortoise is now 1 metre ahead and so on. The argument is that the tortoise is always ahead and so Achilles never catches up. But of course Achilles catches up at the point $100+10+1+0.1+0.01+ \ldots = 111.111\ldots = 111^{1}/_{9}$ metres from the starting point. In the paradox the ratio of the speeds is 10 rather than 2 as in the paper sizes but if Achilles had arthritis and the tortoise was on drugs then maybe Achilles might only have twice the speed of the tortoise.

The golden ratio

ANOTHER INTERESTING WAY to subdivide a sheet of paper is to remove a square from one end. If the remaining piece has the same shape as the original then we say that it is a golden rectangle and the ratio of the sides is the golden ratio. In other words a large golden rectangle can be subdivided into a square plus a small golden rectangle.

Suppose that the piece left after the square is removed has width 1 and length t so that the ratio of the sides is t. Then the square cut off has side t so that the length of the original piece must have been $t+1$ as shown in the diagram opposite.

1 The Golden Number

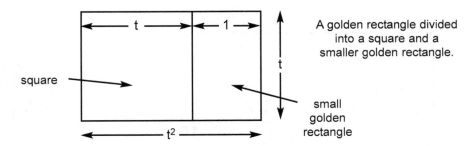

A golden rectangle divided into a square and a smaller golden rectangle.

However, since the original piece has the same shape and its width is t it follows that its length must also have the value t^2. So we have

$$t^2 = t + 1 \quad \text{from which} \quad t^2 - t - 1 = 0$$

Using the familiar formula for the quadratic equation $ax^2 + bx + c = 0$ which is

$$x = \frac{-b \pm \sqrt{(b^2 - 4ac)}}{2a}$$

gives $t = (1 + \sqrt{5})/2 = 1.6180...$. This is the so-called golden number and is often denoted by the Greek letter τ (*tau*). The equation $\tau^2 = \tau + 1$ can be rearranged to give the following relations which should also be noted

$$\tau = 1 + 1/\tau \quad \text{and} \quad 1 = 1/\tau + 1/\tau^2 \quad \text{and so}$$

$1/\tau = 0.6180...$
$\tau = 1.6180...$
$\tau^2 = 2.6180...$

A piece of paper in the golden ratio is rather longer and narrower than A4. For comparison, the old Foolscap size is 13" by 8" which gives a ratio of 1.625 and so is very close to the golden ratio.

We can now see that a golden rectangle can be divided into an infinite number of squares of diminishing size. The arrangement shown below mirrors that of the division of the sheet of A0. Later in this lecture we shall see ways of

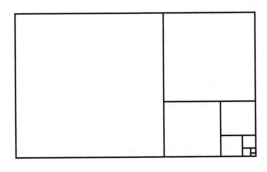

A golden rectangle subdivided into an infinite number of squares.

4 Gems of Geometry

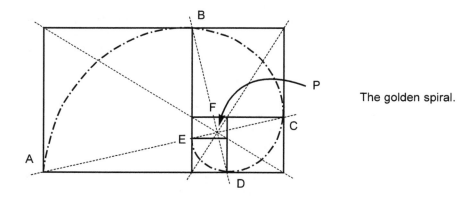

The golden spiral.

dividing a rectangle into a finite number of squares all of which have a different size.

An alternative way to do the subdivision is in a spiral form where the squares converge onto a point P inside the original golden rectangle as shown above. Note how the lines which join the corner points all pass through the point P; these four lines are at 90° to each other in pairs. Thus *AEC* is at right angles to *BFD*. In fact the whole picture simply consists of the basic square repeatedly rotated about the point P by 90° and shrunk by a factor of τ.

A spiral curve can be drawn through the points A, B, C, D, E, F, etc. In a single rotation it changes size by a factor of τ^4 which is about 6.85. This spiral is often called the golden spiral. At first glance it looks as if it is tangential to the squares but in fact it cuts the sides at about 4.75°. In general shape it approximates to that exhibited by many shells although it spirals too quickly.

It has been claimed that the golden ratio is the ideal shape for a room. Most modern houses have rooms that are too thin for their length thus giving an awkward end space. Pictures are often of the golden ratio as well. It is interesting that artists rarely paint a square picture. It is either landscape or portrait in format and typically around the golden ratio.

Fibonacci's rabbits

IN 1202, LEONARDO OF PISA, who was nicknamed Fibonacci (son of good nature), completed a book entitled *Liber Abbaci* (roughly, a Book of Counting). One of the topics concerns the breeding of rabbits. Suppose each pair of rabbits produces another pair of rabbits every month except in their first month of life and that we start with a single pair of new-born rabbits.

In month 1 we have just the original pair and at the start of month 2 we still have just the original pair since they only breed in that month. But at the start of the third month we have 2 pairs and by the fourth month 3 pairs. By the fifth

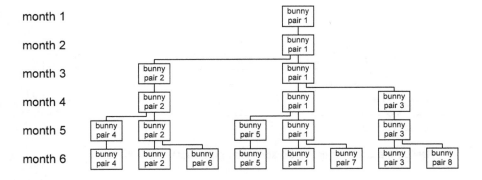

month not only do the original rabbits breed but their first born pair also breed so then we have 5 pairs. By the sixth month the second litter also breed giving 8 pairs. It then really gets going. This is shown in the tree of rabbits above.

A little thought shows that in each month the number of rabbits equals the sum of the numbers in the previous two months. (That's because all the rabbits of the previous month will still be around and all the rabbits of the month before will have bred a new pair.) So we get the famous series of Fibonacci numbers in which each one is the sum of the previous two:

1, 1, 2, 3, 5, 8, 13, 21, 34, 55, 89, ...

As the series progresses the ratio between successive numbers rapidly approaches a limit thus

1, 2, 1.5, 1.666..., 1.600, 1.625, 1.615..., 1.619..., 1.617...

The number the ratio approaches is, surprise, surprise, the golden number, τ. That this should be the case is easily seen since in the limit if each number is x times its predecessor then any group of three successive numbers will have values n, xn, $x^2 n$ and since moreover each number is the sum of its two predecessors it follows that

$x^2 n = xn + n$ and so $x^2 = x + 1$ which is one of the formulae for τ

An interesting property of Fibonacci's numbers is that the square of any one of them differs from the product of the two adjacent numbers by exactly 1, for example

$3 \times 3 = 9 = 2 \times 5 - 1$
$5 \times 5 = 25 = 3 \times 8 + 1$
$8 \times 8 = 64 = 5 \times 13 - 1$
$13 \times 13 = 169 = 8 \times 21 + 1$

Note how the squares are alternately one more than and one less than the product of the adjacent numbers.

6   Gems of Geometry

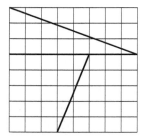 

Dividing a chessboard of 64 squares into a rectangle with 65 squares.

This property is the basis of a very old puzzle. Consider a chessboard and cut it as shown above. It then seems as if the pieces can be rearranged to form a rectangle of size 5 by 13 which has 65 squares whereas the original chessboard had 64 squares.

The question is where did the extra unit of area come from? The answer is that the pieces don't quite fit together and in fact there is a thin parallelogram gap in the rearranged shape. But it is not easy to spot this unless the cutting is done carefully. This puzzle appeared in the Boy's Own Annual for 1917–1918 but surprisingly there was no explanation of where the extra square came from; the discussion implied that it truly appeared from nowhere. Similar puzzles can be made up with the smaller numbers but the gap is then much more obvious.

A related puzzle which is perhaps more surprising is shown below. The various shapes making up the left diagram can be rearranged to give the right diagram. This seems to produce a hole from nowhere!

The reason is that the angles of the two triangles are slightly different. The tangent of the angle at the base of the larger one is 3/8 whereas that at the base of the smaller one is 2/5. These are both the ratios of successive Fibonacci numbers. The angles are about 20.56 and 21.80 degrees respectively and the difference of just over one degree is easily overlooked by the human eye. The top of the left diagram is slightly concave whereas that of the right is slightly convex. This same small angle is in the thin parallelogram in the puzzle above.

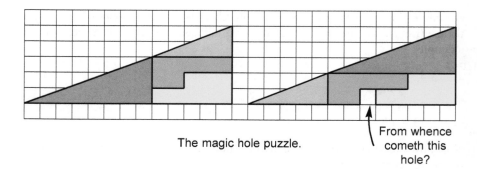

The magic hole puzzle.   From whence cometh this hole?

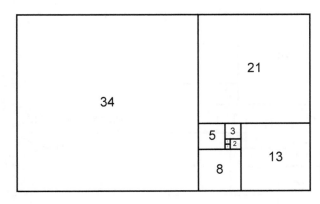

A rectangle of sides 34 × 55 divided into squares whose sides are the first nine Fibonacci numbers.

The Fibonacci numbers provide a way of subdividing an appropriate rectangle into a finite number of squares with one square corresponding to each number. The case of subdividing a rectangle 55 by 34 into 9 squares is shown above. But this doesn't quite solve the problem of dividing a rectangle into a number of *different* squares because the first two Fibonacci numbers are both equal to 1.

However, it is possible to subdivide a rectangle into 9 different squares of sizes 1, 4, 7, 8, 9, 10, 14, 15 and 18. They form a rectangle of sides 32 and 33 as shown below. It makes quite an amusing puzzle to assemble a set of such squares into a rectangle. A set of squares made of aluminium alloy or brass using a unit of perhaps $1/8$" or $1/2$ cm is compact enough to keep in one's pocket and produce at parties to amuse (or bore) your friends. This is not the only subdivision into 9 squares (see Exercise 1 at the end of this lecture for another example) but there are no subdivisions of a rectangle into fewer.

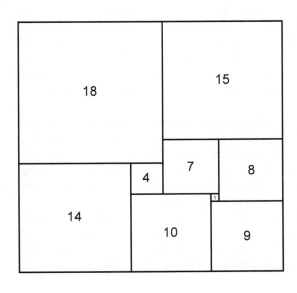

A rectangle of sides 32 × 33 divided into nine different squares.

8   Gems of Geometry

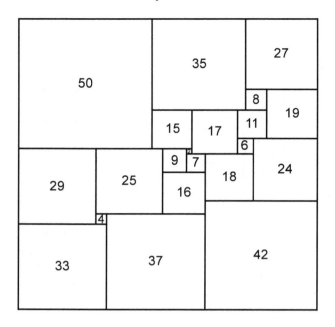

A square of side 112 divided into 21 different squares.

It is also possible to subdivide a *square* into different squares as shown above. The simplest case has 21 different squares of sides 2, 4, 6, 7, 8, 9, 11, 15, 16, 17, 18, 19, 24, 25, 27, 29, 33, 35, 37, 42 and 50. Note that the smallest has side 2 rather than 1. These squares can be arranged to form a square of side 112. I am sure that they would not make a sensible puzzle and would cause one to lose friends quite quickly.

There is a curious relationship between electrical circuits and the subdivision of squares and rectangles into different squares. As a simple example, consider the case of the 32 × 33 rectangle.

Imagine a current flowing from one edge to the opposite edge of a rectangle. The resistance increases as the rectangle gets longer and decreases as it gets wider. These effects cancel out so that all squares have the same resistance irrespective of their size; for simplicity we will assume that the resistance of a square is 1 ohm.

Now suppose that the 32 × 33 rectangle be subdivided into squares and that a uniform current flows from the top edge to the bottom edge as shown opposite. Since the flow is uniform it follows that current never flows from a square to one alongside it but only from a square to one below it. As a consequence the collection of conducting squares is equivalent to a network of 1 ohm resistors connected in this case as shown alongside the rectangle.

Each resistance corresponds to a square and each junction point corresponds to a horizontal division between squares. The current flowing in each resistance is clearly proportional to the width of the corresponding square. Assuming that the current in the resistance corresponding to the square of side 1 is 1 amp it

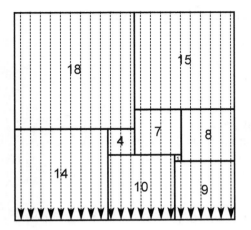 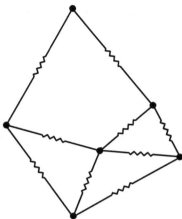

Current flowing from one edge to the other of the 32 × 33 rectangle.

The equivalent electrical circuit of nine equal resistors.

follows that the total current flowing through the network is 33 amps. It is easy to see that the total resistance of the rectangle is 32/33 ohms and that the voltage across the rectangle is 32 volts.

The problem of subdividing a rectangle into different squares thus becomes equivalent to finding an electrical network of 1 ohm resistors in which the current in each resistor is different. A network which exhibits symmetry somewhere (either as a whole or in part of it) will have equal currents somewhere. It is easy to convince oneself by just trying that any network joining two points with less than 9 lines has symmetry somewhere. It follows that 9 different squares is the minimum number for subdividing a rectangle.

The golden ratio and Fibonacci's numbers also turn up in the simplest form of so-called continued fractions which we now briefly explore before returning to the geometrical theme.

Continued fractions

I AM SURE THAT Victorian schoolboys knew all about continued fractions – or maybe they didn't and got beaten as a consequence. Continued fractions are a somewhat neglected topic these days – probably considered too hard to learn at school (high school) and too obscure to learn at university (college). Nevertheless they have intriguing properties which are worth knowing.

Consider any fraction less than 1 such as 7/16. We can write this as

$$\frac{1}{^{16}/_7} = \frac{1}{2^2/_7} = \frac{1}{2 + \frac{1}{^{7}/_2}} = \frac{1}{2 + \frac{1}{3^1/_2}} = \frac{1}{2 + \frac{1}{3 + \frac{1}{2}}}$$

This is usually written in one of various shorthand notations such as

$$\frac{1}{2+} \frac{1}{3+} \frac{1}{2}$$

or even more succinctly as [2, 3, 2]. If a number is greater than 1 such as $4^7/_{16}$ then we can write it as [4; 2, 3, 2].

We recall that some common fractions such as 1/7 expressed in decimal notation do not terminate but exhibit a recurring pattern thus

$$1/7 = 0.142857142857... \text{ sometimes written as } 0.\dot{1}4285\dot{7}$$

All common fractions expressed as continued fractions naturally terminate and of course 1/7 is simply expressed as [7].

A surprising thing about continued fractions is that some numbers which are chaotic as a decimal fraction have a recurring form as a continued fraction. Consider for example the recurring fraction

$$x = [1; 2, 2, 2, ...]$$

This can be written as $1 + y$ where

$$y = \frac{1}{2+} \frac{1}{2+} \frac{1}{2+} ...$$

A moment's thought shows that the fraction following the first 2 is itself y so we have

$$y = \frac{1}{2 + y} \quad \text{from which we get}$$

$$y^2 + 2y - 1 = 0 \quad \text{so that} \quad y = -1 + \sqrt{2}$$

and so finally $x = 1 + y = \sqrt{2} = 1.4142....$

In a similar way we can show that $\sqrt{3} = [1; 1, 2, 1, 2, 1, 2, ...]$. Clearly the most beautiful recurring continued fraction of all has to be [1; 1, 1, 1, ...] and it will come as no surprise that this has the value τ, the golden number.

Some so-called transcendental numbers such as $e = 2.171828...$, the base of natural logarithms, which are a complete mess as a decimal expansion, have a recognizable pattern as a continued fraction

$$e = [2; 1, 2, 1, 1, 4, 1, 1, 6, 1, 1, 8, ...]$$

Others such as π have no recognizable pattern; $\pi = [3; 7, 15, 1, 292, ...]$.

An important property of continued fractions is that if we chop off the sequence then we obtain an approximation to the value. That is not so surprising because if we chop off a recurring decimal such as $0.142857142857...$ then clearly we get an approximation. In the case of continued fractions the chopped off values are known as convergents. The approximations given by a chopped off decimal fraction are always too low but amazingly the convergents of a continued fraction are alternately too low and too high. For example if we consider some successive approximations to $\tau = [1; 1, 1, 1, 1, ...]$, we find

$[1; 1, 1] = 3/2$
$[1; 1, 1, 1] = 5/3$
$[1; 1, 1, 1, 1] = 8/5$

and Lo and Behold these are the ratios of successive pairs of Fibonacci numbers.

An important property of the convergents is that they provide excellent approximations to the original number. In fact it can be shown that a convergent provides the best approximation for the size of its terms. Thus $13/8$ is the best approximation to τ using only ratios involving integers of 13 or less.

The approximation of $22/7$ for π is the first convergent from the continued fraction $[3; 7, 15, 1, 292, ...]$. A convergent is a particularly good approximation if the first term ignored is large. So $[3; 7] = 22/7$ is a good approximation to π because the first term ignored is 15. And $[3; 7, 15, 1]$ is extremely good because the first term ignored is 292.

It is straightforward to compute one convergent from the previous two. Suppose the continued fraction is $[a_0; a_1, a_2, a_3, ...]$ and the convergents are $c_0 = n_0/d_0$, $c_1 = n_1/d_1$, $c_2 = n_2/d_2$ and so on. Then the first two are easy

$c_0 = a_0/1$ so $n_0 = a_0$ and $d_0 = 1$
$c_1 = a_0 + 1/a_1 = (a_1 a_0 + 1)/a_1$ so $n_1 = a_1 n_0 + 1$ and $d_1 = a_1$

The general pattern continues thus

$n_2 = a_2 n_1 + n_0$; $d_2 = a_2 d_1 + d_0$
$n_3 = a_3 n_2 + n_1$; $d_3 = a_3 d_2 + d_1$

The rule is that to get the next convergent you multiply both top and bottom of the previous one by the next term and add them to the top and bottom of the one before that. Let's try that with $\sqrt{2}$ which is $[1; 2, 2, 2, ...]$. The first two

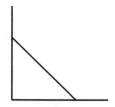

The hypotenuse is √2 times the sides.

convergents are [1] = 1/1 and [1; 2] = 3/2. The next convergent is [1; 2, 2] so the new term added is 2 and then we follow the rule giving the sequence

$$\frac{1}{1} ; \frac{3}{2} ; \frac{2 \times 3 + 1}{2 \times 2 + 1} = \frac{7}{5} ; \frac{2 \times 7 + 3}{2 \times 5 + 2} = \frac{17}{12} ; \frac{2 \times 17 + 7}{2 \times 12 + 5} = \frac{41}{29}$$

So if you want to find a good approximation to a number first convert it to a continued fraction and then choose a convergent using these rules. As an example suppose we are trying to make a right angled corner out of Meccano (or Erector or Marklin) and wish to brace it with an isosceles triangle (that is one with two sides the same). Meccano strips only have holes every 1/2" so we have to choose our bracing with care.

From the right angled triangle shown above (and using Pythagoras' theorem that the square on the hypotenuse is equal to the sum of the squares on the other two sides) it is clear that the length of the brace is going to be √2 times the length of the sides. Now we have just computed the convergents for √2; they are

1/1, 3/2, 7/5, 17/12, 41/29, ...

We can quickly compute the decimal values of these convergents which are

1, 1.5, 1.4, 1.4166..., 1.4137...

whereas the correct value is of course 1.4142...

It's pretty clear that 3/2 is not going to be very good but even 7/5 is only 1% out and 17/12 is more than four times better. In fact the slack in the bolts means that 7/5 will work but of course the slack in the bolts also means it isn't very rigid. So the best choice is 17/12. Remember that the measurements are from the centres of the holes so we need 18 holes = 9" on the diagonal and 13 holes = 6½" along the sides. Any Meccano person will tell you that there is no such thing as a 9" strip so we have to overlap two strips together as shown opposite.

Well that is quite enough about continued fractions so let's get back to some geometry. But before doing so consider the following curious expression

$$x = \sqrt{(1 + \sqrt{(1 + \sqrt{(1 + \sqrt{(1 + \sqrt{(1 + \sqrt{(1 + \sqrt{(1 + \ldots)})})})})})})}$$

A little thought reveals that x must satisfy $x = \sqrt{(1 + x)}$ so that $x^2 = 1 + x$ and hence x is again the golden number.

1 The Golden Number   13

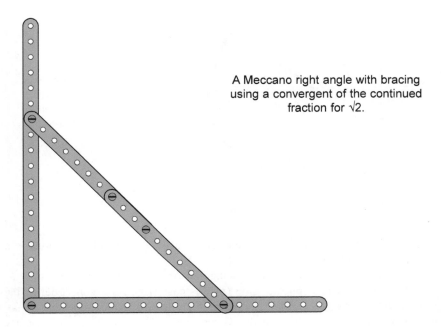

A Meccano right angle with bracing using a convergent of the continued fraction for √2.

Pentagons

THE MOST BEAUTIFUL two-dimensional shape in the world is undoubtedly the regular pentagon. It has long been used as a mystic symbol. If we extend the sides of a pentagon then we get the star-shaped pentagram. If we take a large pentagon and draw its five diagonals then these form a pentagram around a smaller pentagon inside the original pentagon.

The pentagon with its diagonals reveals many instances of the golden number. Suppose for simplicity that the length of the sides of the inner pentagon is 1 unit. Each diagonal of the outer pentagon contains a side of the inner pentagon and two other sections which are sides of an isosceles triangle such as

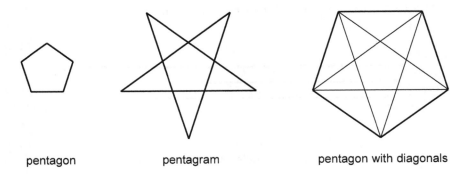

pentagon   pentagram   pentagon with diagonals

14 Gems of Geometry

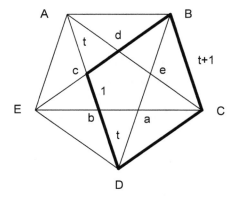

Pentagon with diagonals highlighting internal rhombus.

abD in the diagram above. Suppose the length of these two other sections is *t*. Then the total length of a diagonal is $2t + 1$.

It is important to note that each diagonal of the outer pentagon (such as *EbaC*) is parallel to the opposite side (in this case *AB*). As a consequence the figure *BcDC* is a rhombus which is a parallelogram all of whose sides are equal. The rhombus is shown in bold in the diagram. Now the length of *cD* is clearly $t+1$ and so it follows that the length of a side of the outer pentagon is also $t+1$.

Now notice that the large triangle *ABD* has the same shape as the small triangle *baD*. So the ratios of their sides must be the same. The small triangle has sides in the ratio $t:1$ and the large triangle has sides in the ratio $2t+1:t+1$. And so

$$2t + 1 = t(t + 1) \quad \text{giving} \quad t + 1 = t^2 \quad \text{which is the familiar equation for } \tau.$$

So *t* is actually the golden number τ. The pentagon provides yet another example of the appearance of the golden ratio.

Indeed, the pentagon is absolutely riddled with golden ratios. The inner pentagon has side 1; the larger bits of the diagonal have length τ, the outer pentagon has sides of length τ^2 and the diagonals have overall length τ^3! It is truly a golden figure. These powers of τ are all revealed in the three parts of a diagonal as shown below. Note that since $\tau^2 = \tau + 1$ it follows that $\tau^3 = \tau^2 + \tau = (\tau + 1) + \tau = 2\tau + 1$, the length of the whole diagonal.

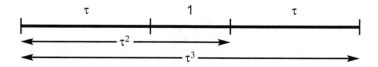

The diagonal is riddled with golden ratios.

1 The Golden Number   15

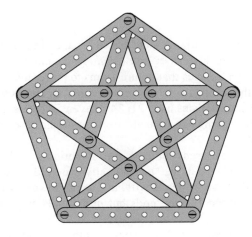

The Meccano pentagon with side 8 and diagonal 13.

It is interesting to compute further powers of τ in the same way. We find

$\tau^4 = 3\tau + 2$
$\tau^5 = 5\tau + 3$
$\tau^6 = 8\tau + 5$

which shows those Fibonacci numbers once more!

Because the ratios of successive Fibonacci numbers are close approximations to τ it follows that a sequence of four Fibonacci numbers forms a good approximation to the series of powers of τ exhibited by the diagonal. We can make a Meccano model of the pentagonal figure using a series such as 5, 8, 13, 21 and the slack in the bolt holes takes up the small errors. We can even do it with 3, 5, 8, 13 as illustrated above but the next series down won't work.

The ancients knew all about the pentagon and the golden ratio. Euclid refers to the fact that the diagonals of a pentagon divide each other in "extreme and mean ratio". The definition in Book VI says "Ακρον και μεσον λογον ευθεια τετμησθαι, οταν η ως η ολη προς το μειζον τμημα, ουτως το μειζον προσ το ελαττον." which obviously means "A straight line is divided in extreme and mean ratio when, as the whole line is to the greater segment, so is the greater to the less." Thus in the diagram the length of the whole line is τ^3, the greater segment has length τ^2, and the lesser segment has length τ and therefore clearly both ratios are τ.

Phyllotaxis

PHYLLOTAXIS MEANS the arrangement of leaves and similar botanical items. It derives from the Greek word φυλον (*phulon*), meaning a leaf.

It has been known for ages that nature seems to prefer the Fibonacci numbers when it comes to the arrangement of branches, leaves and petals. Thus a pansy has five petals and some daisies have 21, 34 or even 55 petals. Flowers that seem to break this rule by having say 6 petals are seen on closer examination to have 2 groups of 3 petals.

However, these favoured numbers are sometimes violated. Thus clover usually has three leaves but just occasionally we find a four-leaved clover. Daisies which mostly have 21 petals will have some flowers with 20 or 22 petals. Poppy heads have around 13 seed pods but individuals may have 11, 12, 14 or even 15.

Arrangements involving Fibonacci numbers also occur as double interlocking spirals in objects such as pineapples and sunflower heads. A pineapple has its sections arranged so that they appear to form 5 spirals in one direction and 8 spirals in the other. Sunflower heads often have 55 and 89 spirals. Typical pinecones have three and five spirals.

The cactus shown below looks somewhat like a pineapple from the side and shows spiky protuberances arranged in spirals of 21 and 34. On top it shows the new buds growing rather like the centre of a sunflower.

Until recently this was all thought to be "in the genes" but now it appears that maybe it is a phenomenon caused by natural processes where the emerging

Two views of the cactus, *Mammillaria tlalocii*. The side shows buds in spirals of 21 and 34. The top is similar to the centre of a sunflower.

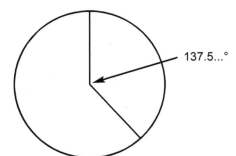

A circle divided in extreme and mean ratio showing the golden angle.

units appear to spread out in a way which minimizes interference and gives each maximum space for growth.

Growth in any plant commences at a conical tip from which new material emerges. The cone might be very sharp as in the tip of a bamboo shoot or quite flat as in a growing sunflower head. It seems however that there is a natural tendency for each new primordial growing bud to come out at an angle of about 137.5° from the previous one. Why this strange angle? Well this angle divides a circle into "extreme and mean ratio" in the words of Euclid. Now we might remember that

$$1 = 1/\tau + 1/\tau^2$$

so that the two angles which divide 360° in extreme and mean ratio are $360/\tau$ and $360/\tau^2$. These angles are 222.492... and 137.508... degrees respectively. The latter is often called the golden angle.

Now if successive buds grow out from the centre each 137.5° from the previous one it turns out that they each have maximum space around them. The diagram below shows ten buds having grown out in sequence at the same rate. Clearly if they were at an angle such as 120° from each other then the fourth bud would find itself up against the first one and the pattern would consist of three lines of buds with wasted spaces between. Indeed we have to avoid any angle which is a rational fraction of 360° (that is a factor of some multiple of 360°). Numbers such as square roots are satisfactory so we might try $360/\sqrt{5}$ which

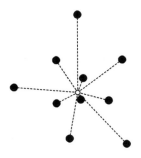

Pattern for τ.

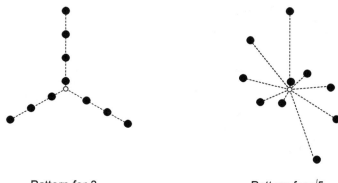

Pattern for 3.   Pattern for √5.

gives 160.996... degrees. But, as seen in the diagram above, it is not that good since a pattern of spiral arms emerges and space is wasted between the arms. A better choice is 360/√2 which gives an angle of 105.442 degrees. But this is not quite so good as τ. The reason that τ is best is related to the fact that its continued fraction has every term equal to one thus [1; 1, 1, 1, ...]. The smaller the terms in the continued fraction the more uniformly distributed is the pattern. The continued fraction for √5 is [2; 4, 4, 4, ...] and that for √2 is [1; 2, 2, 2, ...]. The diagrams above show the distribution for the three cases corresponding to dividing the circle by τ, 3 and √5.

Assuming that the buds remain the same size, it is more realistic to slow their radial progress as they move away from the centre so that they continue to have the same space. Accordingly, a reasonable model is one where the distance out from the centre varies as the square root of the age of the bud. The resulting distribution for the first fifty buds shows that they align themselves as spirals with 13 in one direction and 8 in the other. However, as more buds are added it becomes clear that spirals corresponding to all the Fibonacci numbers can be traced. Adding more buds concentrates the eye on the outer regions where the spirals corresponding to the higher Fibonacci numbers are evident.

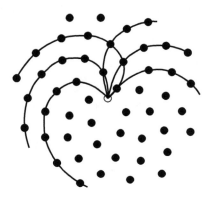

The first fifty buds and the spirals of 8 and 13.

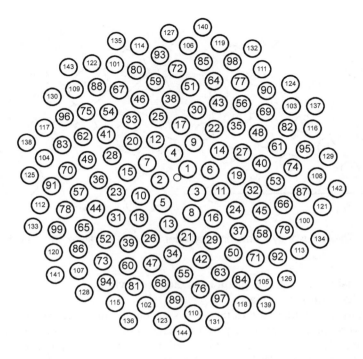

The first 144 buds showing the various spirals.

Thus in the diagram above where the first 144 buds are shown the spirals most evident are those for 13 and 21 although the 34 spiral is beginning to appear. Note that the 21 spirals are formed by the numbers that are 21 apart and so on. The spirals for adjacent Fibonacci numbers always go in opposite directions. We can still trace the lower spirals but the buds on them become further and further apart. Note how the buds corresponding to the Fibonacci numbers themselves are aligned about the vertical. The pattern is clearly the sort of thing we see in a Sunflower head where typically the spirals are in the region of 55 and 89.

We will now consider a rather different model more appropriate to pineapples and fircones which have a general cylindrical appearance. We will suppose that buds are created along the base of a cylinder and migrate upwards at a constant rate. Again we will suppose that the buds emerge at the golden angle of 137.5 degrees apart. A big difference from the flat sunflower is that we can assume that the buds continue to travel at a uniform rate since they will then continue to have the same space about them. The general effect for the first twenty buds will be as in the diagram overleaf where the cylinder has been unrolled. This shows clearly the three spirals up to the right and five to the left and is a good representation of the pattern on a fircone. The spiral of two is also

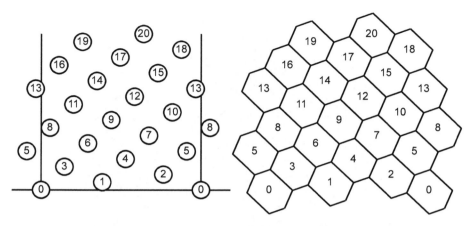

The first twenty points on a cylinder showing spirals of 2, 3 and 5.

clearly visible on the flat picture but not so obvious on a fircone because it swiftly disappears from view.

In this rectangular pattern the spirals which dominate are determined by the ratio of vertical to horizontal separation. A neat presentation is to show the region consisting of the area containing those points closest to each point of the pattern. This is known as the Dirichlet region after the Prussian mathematician Peter Dirichlet (1805–1859). The region is usually hexagonal and is shown in the pattern on the right above which has the same spacing as that on the left and explicitly reveals that the regions which touch correspond to the spirals of 2, 3 and 5.

If we change the ratio of horizontal to vertical separation then the hexagonal regions change shape and the neighbouring regions eventually become different. At the transition positions the regions become rectangular as shown below which illustrates spirals of just 3 and 5. Note that we can still see the spiral of 2 and the nascent spiral of 8 lurking in the pattern.

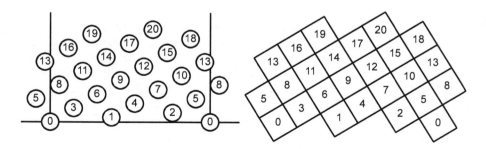

The transition point where the regions are rectangular.

1 The Golden Number   21

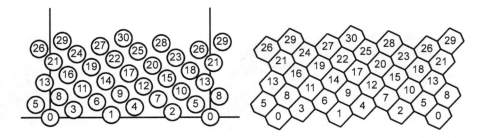

The first thirty points on a cylinder showing spirals of 3, 5 and 8.

Changing the ratio a bit more produces the next stage as shown above when the spirals are of 3, 5 and 8.

Finally, the pattern below illustrates the case of 5, 8 and 13 spirals which corresponds very closely to the structure of a pineapple – although some pineapples may twist the other way.

We can also change the horizontal and vertical separation the other way which corresponds to very rapid growth and reveals spirals of 1, 2 and 3 and then 1, 1 and 2. Remember that the first two Fibonacci numbers are both 1. Finally, we can obtain spirals of 0, 1 and 1. The spiral of 0 occurs when two regions meet each other in both directions.

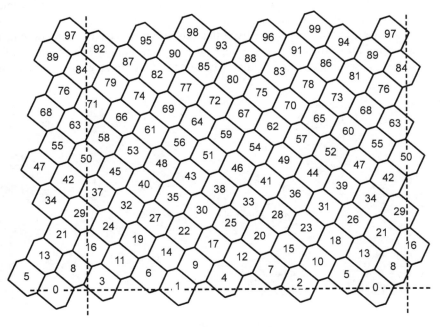

The pattern on a pineapple with 5, 8 and 13 spirals.

22   Gems of Geometry

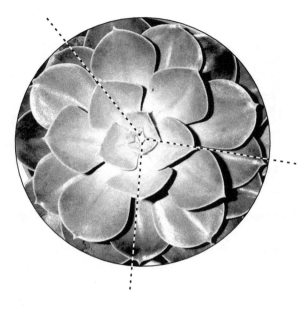

The succulent, *Echevaria elegans*.

Three successive leaves are marked showing the golden angles between them.

Another interesting example of the appearance of the golden number is provided by the succulent *Echevaria elegans* shown above. Successive fleshy leaves grow out at golden angles from each other.

We now return to the question of why it is that the number of petals on many flowers are Fibonacci numbers. Again it seems to concern the emission of buds around a central axis with the golden angle of 137.5° between them.

Suppose that buds grow out from the centre but unlike the sunflower head do not move away from the centre but just grow into petals. The picture on the left below shows the first five petals. Note that the first one to grow is in fact that

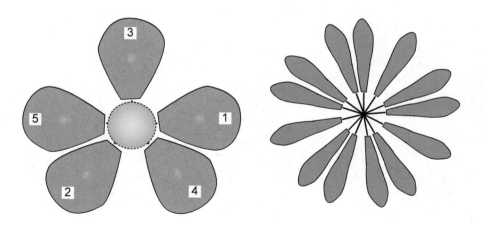

Flowers with 5 and 13 petals.

1 The Golden Number 23

A flower of *Echnicacea purpurea*.
It has 13 petals which show strong indications of being grouped into pairs and singletons.

on the right and the top petal is the third; we have shown it rotated to reveal the symmetry. The sixth petal would be located between the first and third but clearly there isn't room without overlapping. So maybe that is why many plants stop there.

Of course, if the petals are narrower then more can be placed as in the picture on the right which has 13 petals. Amazingly, if we forbid overlapping it is always the case that the first one that will not fit is that immediately after a Fibonacci number of petals have been placed. We also note that the petals are grouped into pairs and singletons. The picture above shows an excellent specimen of *Echnicacea purpurea* which has 13 petals in which some grouping into pairs is evident.

It is fairly easy to show that, as they are added, the nearest petal to the one which is the Nth after a particular Fibonacci number is in fact petal number N. Thus the nearest petal to number 10 is number 2 (since 10 is 2 more than the

Fibonacci number 8) and the nearest to 11 is number 3. Moreover, the spacing depends only on the previous Fibonacci number and decreases by a factor of τ when the next Fibonacci number is reached. Thus petals 9 to 13 are all about 20° from petals 1 to 5 respectively whereas petals 14 to 21 are all about 12.4° from petals 1 to 8 respectively and so on.

Another intriguing feature is that the petals are indeed grouped into pairs and singletons. If there are exactly F_n petals where F_n is the nth Fibonacci number then there are F_{n-2} pairs and F_{n-3} singles. Thus a flower with 13 petals has them grouped into 5 pairs and 3 singletons. Note that this illustrates the fact that for any Fibonacci number

$$F_n = 2F_{n-2} + F_{n-3}$$

For example, $13 = 2 \times 5 + 3$ and $55 = 2 \times 21 + 13$.

In practice of course, nature is not always so precise and especially with the larger numbers of petals some variation is found. Nevertheless, it seems that the golden angle is at the bottom of it all.

The sunflower features in the painting entitled *Virgin of Guadalupe* by Salvador Dali. The head of the Virgin is depicted against a giant sunflower head made of stones with the stones arranged in spirals.

Further reading

CHAPTER 11 of *Introduction to Geometry* by Coxeter covers the key aspects of the golden number and gives for example the mathematical details of Fibonacci's numbers and the pineapple. Pictures of the divisions of squares into squares will be found in *Tilings and Patterns* by Grünbaum and Shephard and also in *Mathematical Snapshots* by Steinhaus although at the time of writing of the latter the decomposition of the square of side 21 was not known; the relationship with electrical circuits is discussed in detail in *Graph Theory* by Bollabás. There is a good discussion on continued fractions in *The Higher Arithmetic* by H Davenport. *The Book of Numbers* by Conway and Guy covers all sorts of numbers and series and includes material on the Fibonacci numbers and phyllotaxis.

A rather different reference work is *Stamping Through Mathematics* by Robin Wilson. This illustrates almost 400 postage stamps depicting various aspects of mathematics. Those of relevance to this lecture include the golden spiral on an 80 cent Swiss stamp of 1987, the decomposition of a rectangle into eleven squares on a 110 pfennig German stamp of 1998, and a somewhat symbolic pineapple on an Israeli stamp of 1961.

An intriguing reference must surely be *De Divina Proportione* by Paccioli published in 1509 and illustrated with drawings of models by Leonardo da Vinci. Curiously, it is written in old Italian rather than Latin. He describes thirteen properties of the golden number. The ninth is that the diagonals of a pentagon intersect in the golden ratio. The original text (in a modern font) begins

> SE nel cerchio se formi el pentagono equilatero e ali suoi doi ₚppinqui anguli se subtéda doi linee recte mosse dali termini deli suoi lati de necessita quelle fra loro se divideráno secondo la nostra ₚpportióe. E cadauna dele lor magior parti semṕ sira el lato del dicto pétagono.

I am grateful to the British Library for permission to quote the above fragment and to Frank Bott of Aberystwyth for providing the following translation.

> IF an equilateral pentagon be inscribed in the circle and to two of its adjacent vertices two straight lines be subtended to the end of the sides, of necessity they divide each other according to our ratio. And each of their larger parts will always be equal to the side of the said pentagon.

It is unclear why the pentagon needed to be inscribed in a circle but the associated diagram shown below certainly helps with the translation of the rest of it which considers an example where the diagonal of the pentagon is 10 units and then CF is $\sqrt{125} - 5$, and AF is $15 - \sqrt{125}$.

We will encounter other properties of the golden number in later lectures.

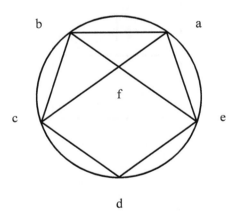

Woodcut by Leonardo da Vinci.

Exercises

1. Find the sizes of the squares in the subdivision of a rectangle shown below. They are all different whole numbers and the smallest has side 2. The diagram is distorted so you can't cheat by measuring!

2. Draw a flower with eight petals and number them in a similar way to the numbering on the flower with five petals in the notes.

3. The continued fraction for π is [3; 7, 15, 1, 292, ...]. We saw that the first convergent [3; 7] is 22/7 which is a frequently used approximation. Compute the next two approximations [3: 7, 15] and [3; 7, 15, 1].

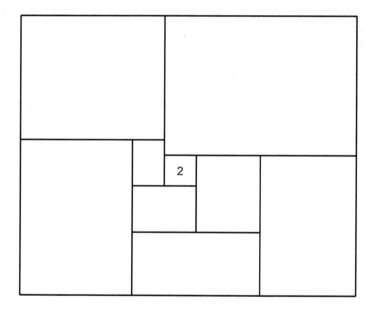

2 Shapes and Solids

THIS LECTURE is about the variety of regular shapes in two and three dimensions. We start by considering the regular plane figures such as the triangle, square, and pentagon and how they can be used to form various regular patterns of tiles. We then move into the third dimension and consider the simple solid figures, such as the tetrahedron and cube which were known in classical times. We conclude by looking at some of the many more elaborate and beautiful figures discovered more recently.

Flatland

ABBOTT (using the pseudonym, A Square) wrote his enchanting book *Flatland* in about 1883. I am sure it has given enormous pleasure to countless readers of many generations. I first read it as a schoolboy and reread it a few years ago. It is a story about the inhabitants of a two-dimensional world. One of the intriguing things is that they are visited by a Sphere who intersects their flat world in a Circle. So all the Flatlanders see of the Sphere is the Circle. When the Sphere first approaches Flatland it appears just as a Point which grows into a Circle and then shrinks again as the Sphere goes away. The Flatlanders cannot believe their eyes at this inexplicable phenomenon.

Another aspect of this book is that it is amazingly politically incorrect. The male inhabitants are geometrical figures. The working classes are isosceles triangles and the angle at the apex distinguishes their skills. A very sharp angle indicates a very lowly person indeed. The middle classes are regular figures with rank according to the number of their sides. Thus a Pentagon outranks a Square and so on. Curiously enough, each generation goes up one rank, so the son of a Pentagon is a Hexagon. The governing aristocracy are Circles. Hence the concern at the visit of the Sphere who appears as a dynamically sized aristocrat.

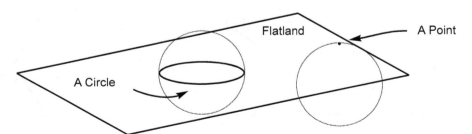

A Sphere intersects Flatland in a Point and a Circle.

The really unacceptable thing these days is that all females are simply two-sided figures with little breadth at all. At one end is a mouth and at the other a sharp tail. So the book is to a very large extent a satire on the social order of the times.

Abbott also explores the one-dimensional Lineland, a fairly boring place, but not so dull as Pointland with its single inhabitant. He also skirmishes with thoughts of the Fourth Dimension and so shall we in the next lecture.

Polygons

WE ARE ALL FAMILIAR with the regular polygons: triangle, square, pentagon, hexagon and so on. There is clearly an infinite number of them. In the diagram below each is shown with its Schläfli symbol which in the case of these polygons is simply the number of sides in curly brackets. Schläfli (1814–1895) was a Swiss mathematician who spent most of his life in Berne and became famous for his discoveries regarding geometrical figures.

A linguistic interlude might be in order. The word polygon comes from two Greek words, πολυς (*polus*), many and γονος (*gonos*), angled. The individual names such as pentagon use the Greek number prefixes: penta-, hexa-, hepta-, octa-, and so on. English is curious, sometimes numbers are taken from the Greek as here but sometimes from Latin as in quartet, quintet, sextet.

One might think that these regular figures were the end of the story but we can consider another way of looking at their construction. We can build a polygon by starting from a point, drawing a line of unit length, turning through an angle and then drawing another line and keep on repeating the process. If the angle we turn through is an exact factor of 360° (such as 120°, 90°, 72° or 60°) then we get one of the normal polygons (triangle, square, pentagon or hexagon respectively).

However, suppose we turn through an angle which is not a factor of 360° but is nevertheless a factor of some multiple of 360°. As an example suppose we turn through 108° which is 1/10 of 1080°, thrice 360°. Then after drawing ten lines and thus making nine turns we do return to the starting point but only after

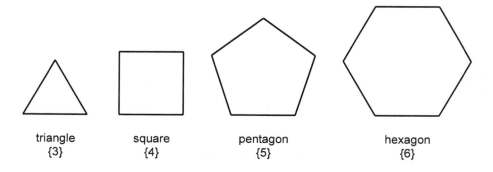

triangle   square   pentagon   hexagon
{3}   {4}   {5}   {6}

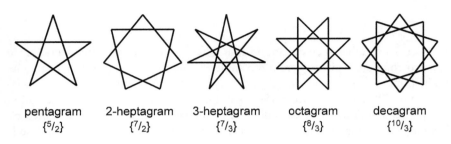

| pentagram | 2-heptagram | 3-heptagram | octagram | decagram |
| {5/2} | {7/2} | {7/3} | {8/3} | {10/3} |

crossing over the lines and thus forming a star shape (a decagram) which is quite regular in form. A number of possibilities are shown above.

The first one is our old friend the pentagram. It has the symbol {5/2} because it has 5 lines and goes around twice. The others have similar symbols. Note that there are two different heptagrams, one goes around twice and the other thrice; moreover, the configuration at the core of the 3-heptagram is in fact a 2-heptagram and the figure at the core of that is an ordinary heptagon.

The concept of density is interesting and will turn up later. It is the number of times a ray starting at the centre of a polygram and going out intersects an edge. The density is of course the second number in the symbol. Thus the density of a decagram is 3. It is clearly a measure of the degree of convolution of a figure.

As we saw in the previous lecture, the pentagram is interesting since it is riddled with golden numbers and in particular we recall that the diagonals cut each other in golden ratios. We might wonder whether the other polygrams have similar romantic properties exhibiting special numbers. Alas, it appears not. Indeed I think this is a typical property of these sorts of mathematical objects – as they get more complex they lose their beauty.

Nevertheless, dedicated readers with an evening to spare might like to try their hands at calculating the ratios in which the diagonals of the 2-heptagram cut each other. Similar techniques to those used in Lecture 1 with the pentagram will do the trick although a lot more effort is required. In fact, we find that if the diagonals are cut in ratios $1:x:1$ where x is the length of the side of the inner heptagon, then x satisfies the quartic equation overleaf.

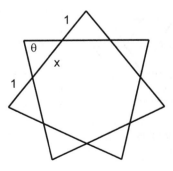

The 2-heptagram showing angle θ, inner heptagon of side x and diagonals of length 2 + x.

$$x^4 + 2x^3 - x^2 - 3x - 1 = 0$$

It is well known that such equations do not have nice solutions. Incidentally, we can also compute x directly from the angle θ which is $77^1/_7°$ (a pretty angle one might think for a seven-sided figure but it is simply a coincidence that one-seventh of 720 subtracted from 180 should have so many 7s in its representation as a proper fraction). Now consider one of the right angled triangles formed by dropping the perpendicular from the corner onto the diagonal. The hypotenuse is 1 and the base is $x/2$, so we get

$$x = 2 \sin \theta/2 = 1.2469...\text{ which is not an interesting number.}$$

We will leave the polygrams at this point. Apart from the pentagram, they are not very exciting and the real point of the discussion has been to show that they can be considered as truly regular figures. We will find that the pentagram, octagram and decagram occur once more when we consider regular figures in three dimensions.

Tiling

IT IS POSSIBLE to cover a flat plane in a completely regular manner with triangles, squares and hexagons but not with any other regular polygon. This is easily seen because for a regular pattern each point must be surrounded by a number of the polygons and so their angle must be a factor of 360°. So we can have six equilateral triangles with angle 60°, four squares with angle 90°, or three hexagons with angle 120° and nothing else.

Underneath each pattern in the diagram below is shown its Schläfli symbol which consists of the symbol for the tile followed by the number of them around each point. Tiling with squares is commonplace in the kitchen and bathroom. The use of hexagons is familiar from the honeycomb made by bees and is also used in wire netting; it has the property of using least material (wax or wire for a given area or gauge size). The form of carbon known as graphite has planes of atoms

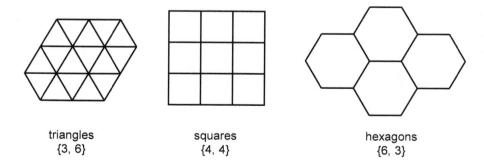

triangles {3, 6}   squares {4, 4}   hexagons {6, 3}

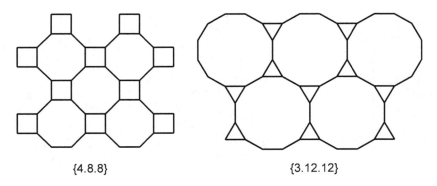

{4.8.8} {3.12.12}

arranged as a hexagonal tiling. The triangle pattern is used for planting orchards since it gives the biggest distance between individual trees.

Note that the tiling of triangles and that of hexagons have the same Schläfli symbol but reversed. This is because they are duals of each other. If we draw a line from the centre of each triangle to the centre of each adjacent triangle then the new lines form the hexagon pattern. Similarly, if we draw a line from the centre of each hexagon to the centre of each adjacent hexagon then the new lines form the triangle pattern. The square tiling is called self-dual for obvious reasons.

If we relax the rule that every tile must have the same shape but instead simply insist that every tile must be a regular polygon and that every point must have the same arrangement of polygons around it then we get a host of further patterns. Four such patterns are shown on this page. The symbol shows the number of sides of the various polygons surrounding each point in order.

The first two patterns are easily obtained from the patterns of all squares and all hexagons by truncation. If we take the square tiles and clip off their corners so that they become octagons then we get the first pattern whereas if we clip the corners off the hexagons we get the second one. The two other patterns are

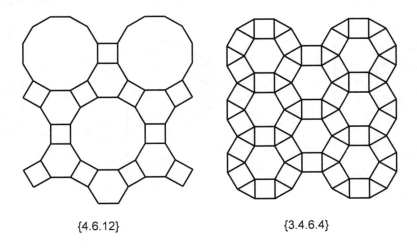

{4.6.12} {3.4.6.4}

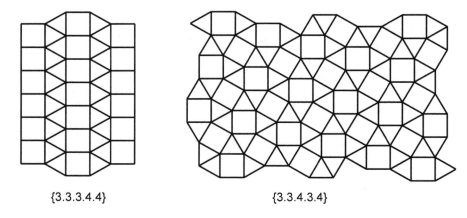

{3.3.3.4.4}   {3.3.4.3.4}

closely related to {3.12.12} and exhibit the twelve-sided theme. The pattern {4.6.12} is perhaps the most intriguing of all the tilings with its mixture of three different polygons which are nevertheless clearly distinguished. The pattern {3.4.6.4} also uses three different polygons but maybe looks a little cluttered; however, it has a simple beauty because it can be looked upon as simply overlapping dodecagons (the Greek prefix for twelve is dodeca-).

Two further patterns can be formed out of triangles and squares. Both have three triangles and two squares at each point but arranged in a different order. That on the left above is curious because it is so obviously different in the two directions whereas the other patterns are not. That on the right looks at first sight like an incoherent mess. It also looks as if it would be different if reflected in a mirror; but this is not so. The reflected version is just the same as the original but slightly rotated.

The final two patterns are shown below. These are formed of hexagons and triangles. The one on the left is unique among the mixed tilings in that each edge has the same status; each edge has a triangle on one side and a hexagon on the

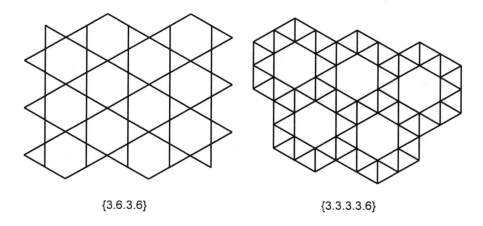

{3.6.3.6}   {3.3.3.3.6}

other. This tiling appears in one of the floors at the Roman villa at Fishbourne in England. The one on the right is interesting because it does become different if reflected in a mirror. So it exists in two forms like right- and left-handed gloves. They are said to be enantiomorphic from the Greek εναντιος (*enantios*), meaning opposite.

The reader might care to try colouring the patterns in order to reveal their structure more clearly. Some appear more attractive than others. Beauty is of course in the eye of the beholder but the human eye likes symmetry and the most attractive patterns seem to be those that exhibit the most symmetries. However, to explain this in more detail would take us outside the scope of these lectures.

Finally, note that the beautiful pentagon does not appear in any of these patterns. We will briefly reconsider this fact at the end of this lecture.

Vision and projection

WE ARE AWARE that the world around us has three dimensions. However, we see everything as two-dimensional images partly because the retina is a two-dimensional surface albeit curved and so the signals sent to our brain only relate to two-dimensional images of the world. Another reason of course is that light travels in straight lines and most objects are not transparent. And so we can only see those parts of things facing towards us.

The fact that we have two eyes enables our brain to provide some information about the distance of objects and so with experience we build knowledge of the three-dimensional world and understand its common shapes.

Because of our need to represent three-dimensional objects on two-dimensional bits of paper we are used to the ideas of projections and this works well with objects with which we are familiar. However, our knowledge is relatively shallow when we consider unfamiliar shapes as we shall see. Projections can take various forms, they can be parallel as if viewed from infinity or they might have a closer viewpoint.

As a simple example consider a cube. Engineering practice is to represent a cube with the front face shown square and then the sides receding at an angle, usually 30°. This has the merit that at least the front face looks correct but in fact this engineering view is not one that can be observed; if we can see part of the side then the front will not appear square. However, it is a familiar and useful representation for many symbolic purposes.

Another possibility is the orthogonal projection. This is the view that would be obtained by looking at it from a long way away (e.g. through binoculars). Of course in order to see anything other than the front face we need to rotate the cube a bit. The second example overleaf shows the effect of rotating the cube a little about the horizontal axis and then by rather more about the vertical axis. (The angles chosen are those in the right angled triangles with sides 7, 24, 25 and

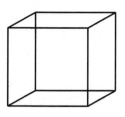 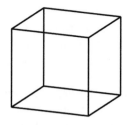 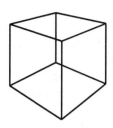

Engineering view of a cube.   Orthogonal view of a cube.   Perspective view of a cube.

5, 12, 13; this gives a reasonable view and results in a position with tidy coordinates.)

The perspective view is obtained by considering a view from a near viewpoint. Parallel lines converge on a point known as the vanishing point. As a consequence the far faces of the cube are shown smaller because they are farther away from the viewpoint. This is the view used in paintings.

We will in general use the orthogonal view because of its mathematical simplicity and the fact that it preserves parallel lines. Another approach is to use two perspective views as seen by each eye and this gives a stereoscopic image. Some examples will be found in Appendices B, D and E.

Five classical polyhedra

WE HAVE SEEN that there are an infinite number of polygons but in fact there are only five regular solids whose faces are regular polygons. This is easily seen by considering the angles at a corner (vertex is the proper term). The sum of the angles of the individual faces which meet at a vertex must be less than 360°. So if the faces are squares (with angles 90°) then the number of faces can only be three which gives the familiar cube. We cannot have less than three anyway and with four the sum of the angles would be exactly 360° which is too much. In fact this gives the tiling of squares which we saw earlier – in a sense the potential solid figure falls flat on its face!

If the faces are triangles (angle 60°) then we can fit together three, four or five at a vertex. In the case of three we get a tetrahedron which has four faces; with four triangles at a vertex we get an octahedron which has eight faces; and with five at a vertex we get an icosahedron which has twenty triangular faces. Attempting six at a vertex gives the triangular tiling.

The last case is where the faces are pentagons (angle 108°) and we can fit together three of these which produces the beautiful dodecahedron with twelve faces.

Faces with more edges are not possible; the next case would be the hexagon (angle 120°) and attempting three of these gives the third regular tiling.

The five figures are shown below together with their Schläfli symbols. These give the symbol for the face followed by the number of faces meeting at a vertex.

The figures show all the edges and sometimes this can be confusing. An alternative is to show the hidden edges as dotted lines or to omit them entirely as overleaf.

The technical term for a solid shape with polygons as faces is a polyhedron from the Greek word ἑδρα (*hedra*), meaning a seat, base or foundation. The plural is polyhedra.

The ancient world was familiar with all these polyhedra and they are often referred to as the Platonic figures. The twelve faced dodecahedron is generally considered the most beautiful and mysterious. It appears in Dali's painting, *In Search of the Fourth Dimension*, and in Escher's lithograph, *Reptiles*. It also appears in *The Last Supper* by Dali in the sense that the room in which the last supper is being held is clearly in the shape of a dodecahedron with pentagonal windows.

Various symmetries should be noted. Apart from the tetrahedron, each face has an opposite face which is parallel to it; similarly each vertex has an opposite vertex. The tetrahedron, however, has a vertex opposite each face and vice versa.

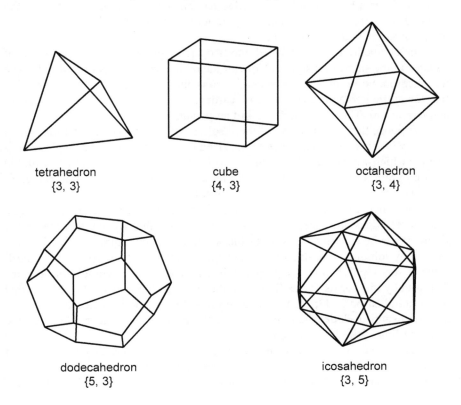

tetrahedron {3, 3}

cube {4, 3}

octahedron {3, 4}

dodecahedron {5, 3}

icosahedron {3, 5}

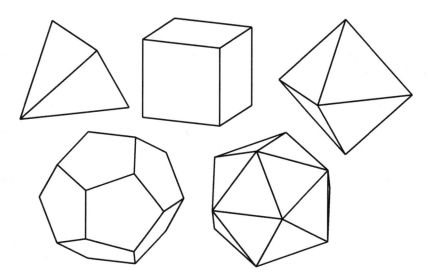

The five Platonic solids with hidden lines removed.

Moreover, for all five solids, each edge has an opposite edge. Apart from the tetrahedron these opposite edges are parallel to one another; in the case of the tetrahedron, they are at right angles.

The octahedron can be divided into two square pyramids and so has a sort of equator in the shape of a square. None of the others has an equatorial divide. The icosahedron, however, has two such divisions and we can think of them as being like the arctic and antarctic circles on earth (or maybe like the tropics of cancer and capricorn). These "circles" are in fact pentagons.

I suppose that the only one of these solids with which we are familiar on an everyday basis is the cube; this is emphasized by the fact that it is indeed the only one for which we have an Anglo-Saxon word. We are of course familiar with the square pyramid of the Pharaohs but the tetrahedron is a pyramid on a triangular base. All five solids occur in crystals or primitive organisms.

If we slice a small corner off a cube, then the revealed section is a triangle since three edges and faces meet at a corner of the cube. We say that the vertex figure of a cube is a triangle. Strictly speaking, the vertex figure is defined as that formed by joining the midpoints of the edges meeting at the vertex. The Schläfli symbol for a polyhedron is properly interpreted as the symbol for its face followed by that for its vertex figure. So the cube has Schläfli symbol $\{4, 3\}$, that of the tetrahedron is $\{3, 3\}$, the octahedron is $\{3, 4\}$, the icosahedron is $\{3, 5\}$ and the beautiful dodecahedron is $\{5, 3\}$. We now see that the symbols for the regular tilings follow the same rule.

Despite our familiarity with the cube it is not obvious that if we slice it diagonally through the centre, then the cross-section is a regular hexagon. We can contemplate this by imagining what the inhabitants of Flatland would see if

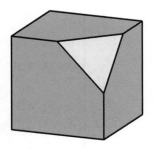 

Two diagonal sections through a cube.

a Cube were to penetrate and cross through their land. The sequence of events as seen by the Flatlanders depends upon how the Cube approaches.

If the cube comes vertex first then it will first appear as a Point and then as a growing Triangle. As the next three vertices of the cube pass through Flatland, the Triangle gets blunted at its three corners and turns into an irregular hexagon. When the cube is exactly halfway the Hexagon becomes regular. The process then repeats in reverse.

If the cube comes face first then it will suddenly appear as a Square and will remain so and then suddenly disappear.

If the cube approaches edge first then it first appears as a Line which immediately turns into a Rectangle whose length is that of the line but whose width gradually grows until it turns into a Square. The Square continues to grow into a Rectangle in the opposite sense until the long edge is $\sqrt{2}$ times the original line at which point the cube is halfway and four of its vertices are in Flatland. The process then repeats in reverse.

So the cube contains within it quite a variety of cross-sections: triangles, rectangles, squares and hexagons. The diagram below shows the appearance of a fleet of three cubes at the different orientations as they pass through Flatland.

We can similarly contemplate what would appear if the other regular solids were to pass through Flatland. The tables overleaf show what happens in the case

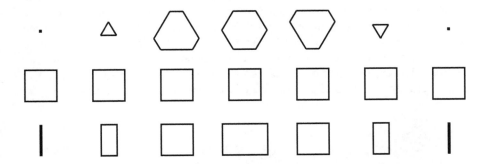

A sequence of views of a fleet of three Cubes of different orientations as seen by the Flatlanders.

Object	Sequence
Tetrahedron	point, triangle
Cube	point, triangle, hexagon, triangle, point
Octahedron	point, square, point
Dodecahedron	point, triangle, hexagon, triangle, point
Icosahedron	point, pentagon, decagon, pentagon, point

Vertex first.

Object	Sequence
Tetrahedron	triangle, point
Cube	square
Octahedron	triangle, hexagon, triangle
Dodecahedron	pentagon, decagon, pentagon
Icosahedron	triangle, 9-gon*, dodecagon*, 9-gon*, triangle

Face first.

Passing through Flatland, vertex and face first. Those marked * are not regular.

of a solid approaching vertex first and face first. Note that the tetrahedron is somewhat different to the others since if it approaches vertex first then it leaves face last and vice versa.

We have already seen that if a cube approaches edge first then the section is a series of rectangles with the section being a square at two positions. Perhaps surprisingly the sections of a tetrahedron passing edge first are also rectangles and at the midway point the section is a square as shown below. The other cases of passing through Flatland edge first are left to the reader.

As a linguistic aside note that the proper name for a 9-gon is an enneagon from the Greek for nine which is εννεα (*ennea*). However, it is often referred to as a nonagon using the Latin prefix.

It is obvious that the angles between the faces of a cube are all 90°. The angles between the faces of a tetrahedron are 70° 32'. (That is 70 degrees plus 32 minutes – remember that there are 60 minutes to a degree.) The angles between the bonds in a carbon atom are 109° 28' which is 180° minus 70° 32'; the atom is at the centre of a tetrahedron and the four bonds are in the direction of the vertices. The angles between the faces of an octahedron are also 109° 28'. The angle between the faces of a solid figure is known as the dihedral angle.

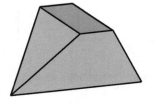 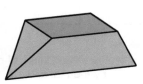

Two sections through a tetrahedron.

Duality

THIS IS a good moment to say a little more about duality which we briefly mentioned when introducing the regular tilings. We might have noted that the Schläfli symbol for the cube and octahedron are related symmetrically. The cube has three faces at each vertex and each face has four sides whereas the octahedron has four faces at each vertex and each face has three sides. The cube has six faces and eight vertices whereas the octahedron has eight faces and six vertices. Both have twelve edges.

Moreover, if we join the midpoints of the six faces of a cube then we obtain an octahedron. And similarly if we join the midpoints of the eight faces of an octahedron then we obtain a cube.

We say that the cube and octahedron are dual figures; the properties of one can be deduced from the other by interchanging the roles of the vertices (points) and the faces (planes). The edges (lines) occupy a central role.

The same dual relationship is exhibited by the dodecahedron and icosahedron. The tetrahedron however is self-dual; if we join the centres of the faces of a tetrahedron then we get another tetrahedron. The diagram below shows these combinations. The internal figure is shown solid whereas the external one is transparent.

It is clear that the cube and octahedron are closely related as are the dodecahedron and icosahedron. What is perhaps remarkable is that all five figures can be found lurking inside or around each other.

Consider a cube and starting from a corner add lines joining it to the diagonally opposite corners of the three faces around the initial corner (see overleaf). Then add lines joining those three corners to each other. The result is that we have a figure composed of six equal lines and four points. Well, the only figure like that is a tetrahedron. A tetrahedron can be drawn in a cube in two different ways. Each joins a different set of four of the eight vertices of the cube. Each has an edge as one diagonal of each of the six faces of the cube.

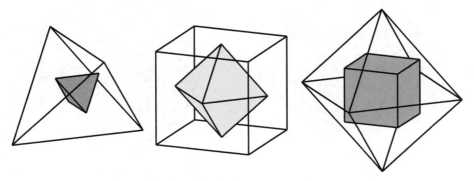

A tetrahedron in a tetrahedron, an octahedron in a cube, a cube in an octahedron.

40 Gems of Geometry

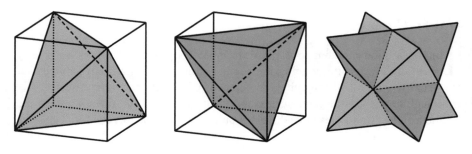

Tetrahedra in a cube and the compound Stella Octangula.

If we take both tetrahedra and remove the cube then we have the simplest so-called compound figure. It was mentioned by Paccioli in his book, *De Divina Proportione*. It was rediscovered by Kepler who called it the Stella Octangula (eight-pointed star). The fact that two tetrahedra are involved reflects the self-dual nature of the tetrahedron.

Another attractive compound is formed from a cube and an octahedron in which they penetrate each other. The edges of one intersect the edges of the other at their midpoints and at right angles. The existence of this compound is another consequence of the duality between the cube and octahedron. There is, of course, a similar compound formed from a dodecahedron and an icosahedron. Escher's mezzotint entitled *Crystal* shows the compound of cube and octahedron and his woodcut entitled *Four Regular Solids* shows both compounds interlaced.

Some of the vital statistics of the five polyhedra are given in the table opposite. Note especially that the formula $F+V-E$ has the same value (2) in each case. This is a famous formula of Euler (1707–1783), the Swiss mathematician. We shall meet it again in Lecture 5 when we discuss some aspects of Topology.

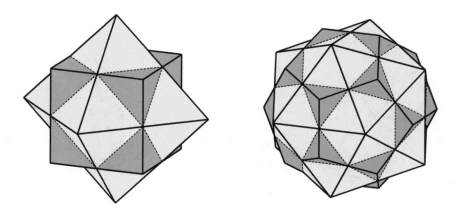

Compounds of the cube and octahedron and the dodecahedron and icosahedron.

Object	Symbol	Faces	Edges	Vertices	F+V–E	Dihedral
Tetrahedron	{3, 3}	4	6	4	2	70° 32'
Cube	{4, 3}	6	12	8	2	90°
Octahedron	{3, 4}	8	12	6	2	109° 28'
Dodecahedron	{5, 3}	12	30	20	2	116° 34'
Icosahedron	{3, 5}	20	30	12	2	138° 11'

Two other rather amazing compounds consist of five interlocking tetrahedra and five interlocking cubes. These exist because of a perhaps surprising relationship between the dodecahedron and the cube. In order to see this, first consider the effect of joining together four vertices in adjacent faces of a dodecahedron as shown in the first diagram below. This results in a four sided figure. Each side is clearly the same length because each is a diagonal of equal sized pentagons. Moreover, the angles are clearly the same since they are formed by adjacent diagonals of two faces in a similar way. So the figure is a square. If we continue to draw lines in adjacent pentagonal faces then eventually we find that we have drawn six squares which are the six faces of a cube inside the dodecahedron as shown in the second diagram.

Each of the 12 edges of the cube is a diagonal of the 12 faces of the dodecahedron. Because a pentagon has five diagonals, it follows that there are five different ways of constructing a cube inside a dodecahedron. The five cubes are oriented as shown below.

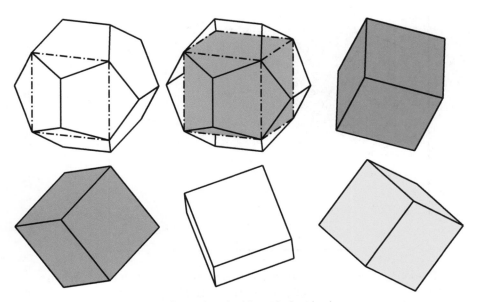

Building five cubes inside a dodecahedron.

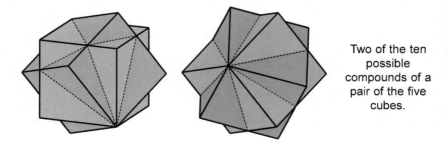

Two of the ten possible compounds of a pair of the five cubes.

If we now remove the dodecahedron we are left with the remarkable compound of five interlocking cubes. Each vertex of the original dodecahedron is shared by two cubes (the dodecahedron has 30 vertices and five cubes have 60 between them) and of course these two cubes also share the opposite vertex. Two cubes can be chosen from five in ten different ways and so there are ten possible compounds of a pair of cubes. Two such pairs are shown above.

The cubes intersect each other at points which divide their edges in the golden ratio; this is because the edges of the cubes are diagonals of the

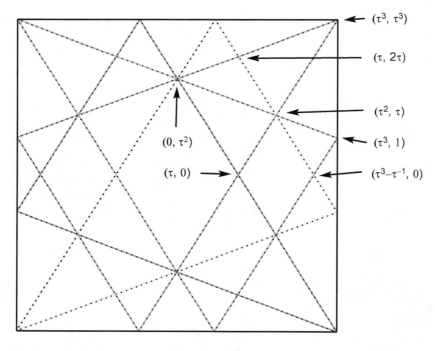

The four coloured cubes intersect a face of the white cube in four coloured triangles. If the cubes have edge of length $2\tau^3$ and the white square is centred at the origin then the coordinates of some points are as shown.

pentagonal faces of the dodecahedron and we saw in the previous lecture that the diagonals of a pentagon intersect each other in the golden ratio.

Although the compound of a pair of cubes is easy to comprehend, the compound of all five cubes is rather intricate. A first step in its understanding can be obtained by considering the face of one cube (the white one for example) and the lines of intersection of the four other cubes with it. This is shown opposite where the intersecting cubes are coloured red, blue, green and yellow and the intersecting lines have the appropriate colour. Note that the cubes intersect each other in triangles so that the figure is just four overlapping triangles; the common area is a rhombus which will be mentioned again later. The cognocenti of golden numbers will delight in the coordinates of the various points of intersection of the triangles which are as shown taking the side of a cube to be $2\tau^3$.

Finally, the diagram below shows a view of the compound of all five cubes looking straight at the white face. The solid lines are the edges of the cubes and form the convex edges of the compound; the dashed lines are lines of intersection of the cubes and are the concave edges of the compound.

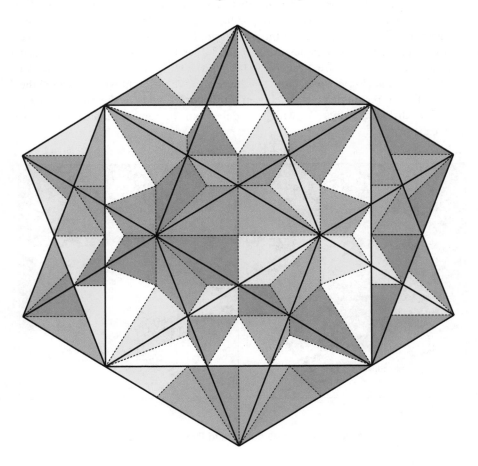

We recall that a tetrahedron can be drawn inside a cube. If we do that for each cube we obtain a compound of five tetrahedra. Moreover, there are two ways to draw a tetrahedron in a cube (left- and right-handed) and consequently there are two different forms of the compound of five tetrahedra which are enantiomorphic.

We could of course put all ten tetrahedra together. This can be looked upon as the result of putting together the two forms of the compound of five tetrahedra or alternatively we might consider the tetrahedra in pairs (recall that such a pair of tetrahedra form a stella octangula) and thus the result is equivalent to a compound of five stellae octangulae.

As well as the figure composed of five cubes, there is a dual consisting of five octahedra surrounding an icosahedron. The five tetrahedra are self-dual.

These remarkable compounds can only really be appreciated by handling actual models. The photographs below show (elderly) paper models of the five cubes and the five tetrahedra.

The icosahedron also has some interesting properties regarding internal plane figures. Two adjacent vertices plus their opposites form the corners of a golden rectangle. See the diagram opposite. An icosahedron has twelve vertices altogether and as a consequence three such golden rectangles define the twelve vertices. These three rectangles are at right angles to each other. This is in fact the thirteenth and last effect regarding the golden number described by Paccioli in his *De Divina Proportione*. Note that there are five different ways in which the group of three rectangles can be chosen – this corresponds to the fact that there are five different ways in which a cube can be placed inside the dual dodecahedron.

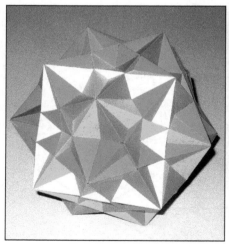 

Models of the five cubes and five tetrahedra.

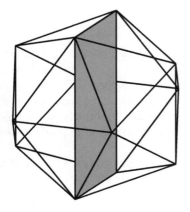 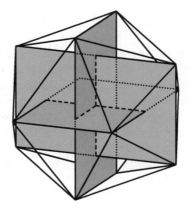

Three mutually orthogonal golden rectangles inside an icosahedron.

It is obvious that groups of five adjacent vertices of an icosahedron form a pentagon since we simply have to consider the five triangular faces meeting at any vertex. There are of course twelve such pentagons one corresponding to each vertex. It is perhaps more surprising that groups of three vertices form an equilateral triangle. The vertices are not adjacent as they are for the pentagons but are the opposite vertices of the three faces surrounding any given face. There are therefore twenty such triangles one corresponding to each face of the icosahedron. Note that the sides of the triangles are not edges of the icosahedron. In both cases the polygons form parallel pairs (and in the case of the triangles are parallel to faces of the icosahedron). Two such parallel pairs are shown below.

We will meet these twelve pentagons and twenty triangles again in a moment.

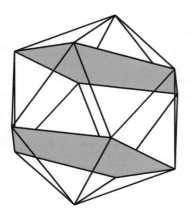 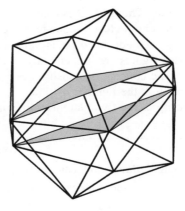

Pairs of parallel pentagons and triangles in an icosahedron.

Kepler and Poinsot

WE IMPLIED ABOVE that the five regular polyhedra (the Platonic figures) were the only regular figures possible. However, just as we considered the polygrams such as the pentagram to be regular polygons then we should admit the possibility that a regular solid might be non-convex and have faces consisting of polygrams. At first sight this seems ludicrous since intuitively one cannot imagine prickly stars neatly fitting together. Nevertheless, it is possible as was shown by Kepler and Poinsot.

Kepler (1571–1630) is best known for his laws of planetary motion such as that the orbits of the planets are ellipses. He showed that there are two regular figures whose faces are all pentagrams. One has five pentagrams at each vertex and the other has three; their Schläfli symbols are $\{5/2, 5\}$ and $\{5/2, 3\}$. The faces naturally intersect but that is reasonable since the sides of the pentagram intersect anyway. They have heavy names, the Small Stellated Dodecahedron and the Great Stellated Dodecahedron respectively. These names reflect that they can be built up from the dodecahedron which lies at their centre. They both have twelve faces arranged parallel to the faces of this central dodecahedron. Curiously, the small stellated dodecahedron is depicted on the floor of St Marco in Venice which predates Kepler by many years; it is attributed to Paolo Uccello.

The duals of these were discovered many years later by Poinsot (1777–1859). They are the Great Dodecahedron $\{5, 5/2\}$ which has twelve pentagonal faces and the Great Icosahedron $\{3, 5/2\}$ which has twenty triangular faces. In both cases the vertex figure is a pentagram and they both have twelve vertices arranged as the vertices of an icosahedron. The faces are the twelve pentagons and twenty triangles inside an icosahedron discussed in the last section.

Of the stellated dodecahedra perhaps the small one with five pentagrams at each vertex is more attractive; the large one is really too prickly to be beautiful. The small stellated dodecahedron appears in Escher's lithographs, *Order and Chaos* (two versions) and *Gravity*. It has also been used as the basis for glass lanterns. The great icosahedron is somewhat confusing; its triangular faces do not stand out clearly since they are penetrated by three vertex figures. However, the great dodecahedron is very attractive; both the pentagonal faces and the vertex figures are immediately obvious.

The concept of density was mentioned with respect to the polygrams. The density is the number of times a ray out from the centre intersects a face. The density of the five Platonic solids is of course one. But the density of these non-

Object	Symbol	Faces	Edges	Vertices	Density
Small stellated dodecahedron	$\{5/2, 5\}$	12	30	12	3
Great stellated dodecahedron	$\{5/2, 3\}$	12	30	20	7
Great dodecahedron	$\{5, 5/2\}$	12	30	12	3
Great icosahedron	$\{3, 5/2\}$	20	30	12	7

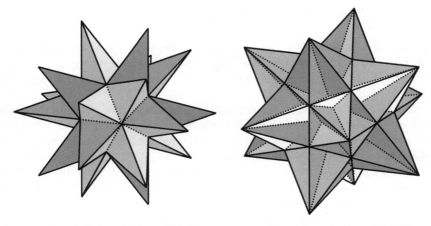

Great stellated dodecahedron {5/2, 3}   Great icosahedron {3, 5/2}
The Kepler–Poinsot figures with density 7.

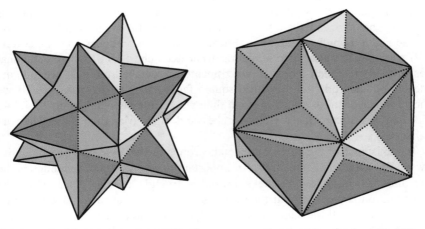

Small stellated dodecahedron {5/2, 5}   Great dodecahedron {5, 5/2}
The Kepler–Poinsot figures with density 3.

convex figures is not one. It is three in the case of the small stellated dodecahedron and its dual and it is seven in the case of the great stellated dodecahedron and its dual. Note that in calculating the density, the penetration of the core of a pentagram counts as two intersections. It is probably not a coincidence that those with lower density are more attractive since the higher density is clearly associated with a greater degree of convolution.

The table shows the statistics for these figures. Note that Euler's formula does not work for two of them. This is because Euler's formula relates to all the individual segments of faces and edges as cut up by the intersections with each other whereas we are considering the faces and edges as a whole.

The Archimedean figures

WHEN CONSIDERING tilings we decided that the key to a regular pattern was that it should be composed of any regular polygons provided that each point should have the same polygons around it. This produced eight additional regular tilings.

We can similarly extend our definition of regular solids to permit the faces to be any regular polygons provided the same arrangement of faces occurs at each vertex. Disallowing intersecting faces for the moment, this introduces the thirteen so-called Archimedean figures.

A number of these are obtained by truncating the five Platonic figures much as we saw how some new tilings could be produced by clipping the corners off the squares and hexagons in the uniform tilings.

Thus if we file down the vertices of a tetrahedron, we get a new triangle at each vertex and the original triangular sides become hexagons. This is known (boringly) as the truncated tetrahedron which we can denote by {3.6.6} in a similar notation used for the tilings. Thus it has 12 vertices, 8 faces (4 hexagons and 4 triangles) and 18 edges. Each vertex has a triangle and two hexagons around it. Some edges have a hexagon either side and some have a hexagon on one side and a triangle on the other. Further truncation produces an octahedron.

If we truncate a cube then we get the truncated cube {3.8.8} with 6 octagonal faces and 8 triangular faces. If we truncate an octahedron then we get the truncated octahedron {4.6.6} with 6 square faces and 8 hexagonal faces. If we truncate either of these a further stage then we get the cuboctahedron {3.4.3.4} which has six square faces (like a cube) and eight triangular faces (like an octahedron).

The existence of the elegant cuboctahedron is further evidence of the duality between the cube and octahedron. An important property is that each edge is the same and has a square on one side and a triangle on the other. Thus the cuboctahedron is more regular than figures such as the truncated cube in which the edges are not all the same. Another property of the cuboctahedron is that it can be looked upon as two halves divided by a hexagon. In fact there are four such equatorial hexagons each containing one side of all six squares.

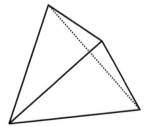 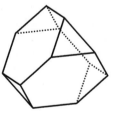

A normal tetrahedron and a truncated tetrahedron {3.6.6}.

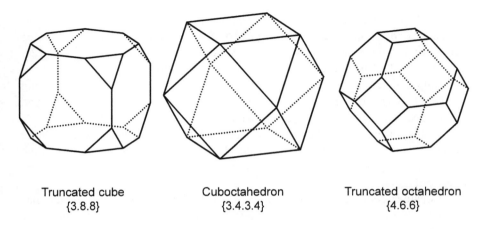

Truncated cube
{3.8.8}

Cuboctahedron
{3.4.3.4}

Truncated octahedron
{4.6.6}

As expected, a similar pattern emerges by truncating the dodecahedron and icosahedron. We have the truncated dodecahedron {3.10.10} and the truncated icosahedron {5.6.6}. If we truncate either of these further then we get the midway icosidodecahedron {3.5.3.5} which has 12 pentagonal faces (like a dodecahedron) and 20 triangular faces (like an icosahedron). The icosidodecahedron is extra regular like the cuboctahedron in that each edge is the same. It also is divided in two by equatorial polygons which in this case are decagons. There are five such equatorial decagons and each contains one of the five sides of every pentagonal face.

Incidentally, the truncated icosahedron with its 12 pentagonal faces and 20 hexagonal faces is the shape of the modern soccer ball. Of deeper interest is perhaps the recent discovery of the strange molecule of carbon, C_{60}, whose 60 atoms are arranged as the 60 vertices of the truncated icosahedron. This allotrope of carbon is sometimes known as Buckminsterfullerene after the architect.

It is interesting to note the close relationship between the group of three cube/octagon based figures and the three dodecahedron/icosahedron based

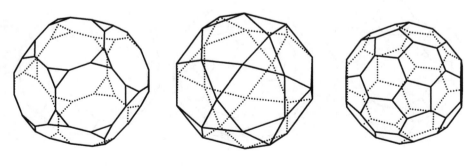

Truncated dodecahedron
{3.10.10}

Icosidodecahedron
{3.5.3.5}

Truncated icosahedron
{5.6.6}

figures. In fact the symbols for the first group have 4s in the places where those of the second group have 5s (the 8 and 10 correspond also). If we go one step further and replace the 5s by 6s then we obtain three tilings namely {3.12.12}, {3.6.3.6} and {6.6.6} the last being of course the uniform tiling of hexagons normally denoted by {6, 3}. Many properties correspond, for example the tiling {3.6.3.6} has the same extra regularity as the cuboctahedron and icosidodecahedron with all edges the same.

So that introduces seven of the thirteen Archimedean figures. The remaining six can be introduced in various ways but a good way is to consider some of the remaining tilings and what happens if we reduce the number of polygons around a key polygon.

Consider for example the tiling {4.6.12} shown below left which has alternate hexagons and squares arranged around a dodecagon. Suppose we reduce the dodecagon to a decagon – this forces the pattern to fold up and produces {4.6.10} which is sometimes called the truncated icosidodecahedron. A further similar reduction produces {4.6.8} the truncated cuboctahedron. (These names are sometimes considered incorrect because straightforward truncation produces rectangles rather than squares at the old vertices and some realignment is then required to obtain the regular figures.)

Two other regular figures are obtained by similarly starting from the tiling {3.4.6.4} which is that formed by overlapping dodecagons. More explicitly, it consists of squares and triangles around a hexagon. If we reduce the hexagon to a pentagon we get {3.4.5.4} which is known as the rhombicosidodecahedron for reasons which need not bother us. Reducing the pentagon to a square produces {3.4.4.4} known as the rhombicuboctahedron.

Both the truncated icosidodecahedron and the rhombicosidodecahedron have 30 square faces and these are oriented in exactly the same way as the 30 faces of the compound of five cubes inscribed inside a dodecahedron which we met earlier.

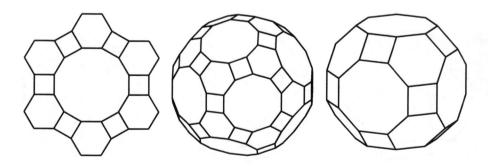

Tiling
{4.6.12}

Truncated icosidodecahedron
{4.6.10}

Truncated cuboctahedron
{4.6.8}

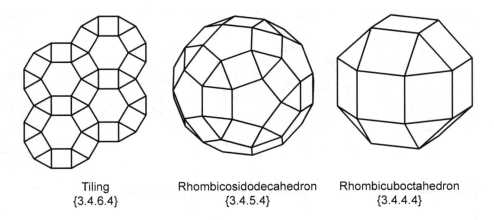

Tiling
{3.4.6.4}

Rhombicosidodecahedron
{3.4.5.4}

Rhombicuboctahedron
{3.4.4.4}

The final two figures are obtained from the tiling {3.3.3.3.6} which it may be remembered came in enantiomorphic forms (like left- and right-handed gloves). Reducing the central hexagon to a pentagon produces {3.3.3.3.5} known as the snub dodecahedron. Reducing the pentagon to a square produces {3.3.3.3.4} known as the snub cube. Like the original tiling both these snub figures come in enantiomorphic forms. Incidentally, the word snub comes from the Old Norske *snubba* meaning to cut short – it's nice to have some Scandinavian influence for a change after all that Latin and Greek. The snub cube has six square faces like a cube and similarly arranged in parallel pairs but somewhat twisted. In addition it has 32 triangular faces. The snub dodecahedron has twelve pentagonal faces like the dodecahedron and in addition has 80 triangular faces.

That concludes the thirteen Archimedean figures. So altogether there are 18 convex figures (5 regular and 13 mixed) and 11 tilings (3 regular and 8 mixed). We have also seen that many of these are related by various transformations between them. These relations are depicted on the next two pages.

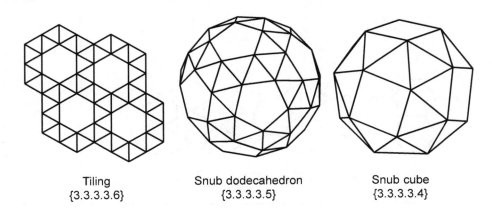

Tiling
{3.3.3.3.6}

Snub dodecahedron
{3.3.3.3.5}

Snub cube
{3.3.3.3.4}

52 Gems of Geometry

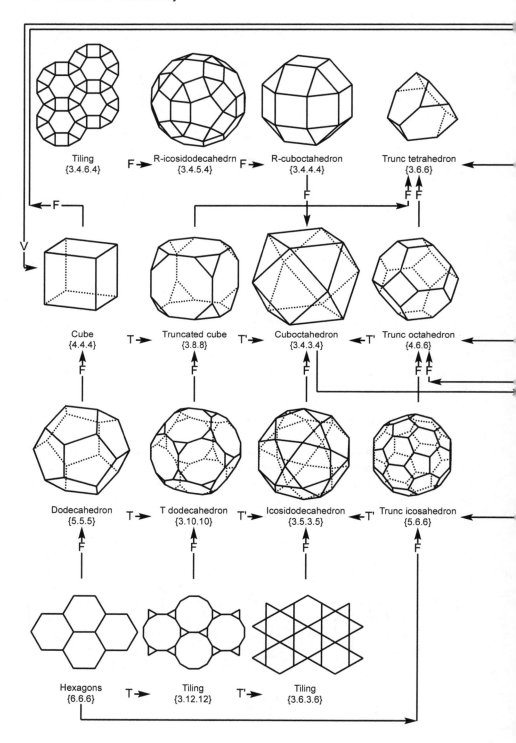

2 Shapes and Solids 53

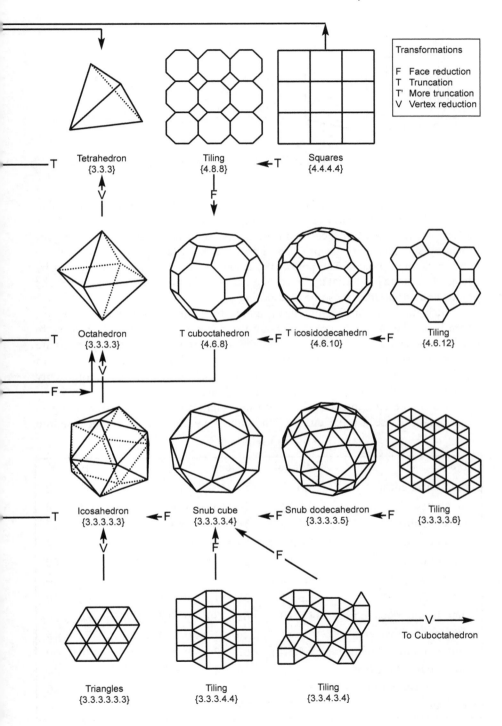

54   Gems of Geometry

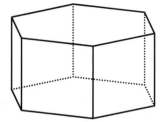 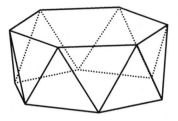

A hexagonal prism and hexagonal antiprism.

There are also the prisms and antiprisms. These satisfy the requirement that a regular figure should have regular polygonal faces and each vertex should have the same arrangement of faces about it. A prism consists of two equal polygons joined by squares. A square prism is simply a cube. Clearly, there is an infinite number of such prisms. The antiprisms consist of pairs of polygons joined by triangles arranged alternately. A triangular antiprism is an octahedron. The prisms and antiprisms are somewhat boring.

The statistics of the Archimedean figures are summarized in the table below. There is another set of figures which are the duals of the Archimedean ones. In the Archimedean figures, although the faces are all regular, the vertex figures are not. The duals on the other hand have regular vertex figures but the faces are not regular. They don't look very attractive presumably because the eye more easily recognizes regular faces than regular vertex figures.

Two of the Archimedean duals are important and are worth describing in some detail. These are the duals of the cuboctahedron and icosidodecahedron

Object	Symbol	Faces	E	V
Trunc tetrahedron	{3.6.6}	8 = 4(3)+4(6)	18	12
Trunc cube	{3.8.8}	14 = 8(3)+6(8)	36	24
Cuboctahedron	{3.4.3.4}	14 = 8(3)+6(4)	24	12
Trunc octahedron	{4.6.6}	14 = 6(4)+8(6)	36	24
Trunc dodecahedron	{3.10.10}	32 = 20(3)+12(10)	90	60
Icosidodecahedron	{3.5.3.5}	32 = 20(3)+12(5)	60	30
Trunc icosahedron	{5.6.6}	32 = 12(5)+20(6)	90	60
Trunc icosidodecahedron	{4.6.10}	62 = 30(4)+20(6)+12(10)	180	120
Trunc cuboctahedron	{4.6.8}	26 = 12(4)+8(6)+6(8)	72	48
Rhombicosidodecahedron	{3.4.5.4}	62 = 20(3)+30(4)+12(5)	120	60
Rhombicuboctahedron	{3.4.4.4}	26 = 8(3)+18(4)	48	24
Snub dodecahedron	{3.3.3.3.5}	92 = 80(3)+12(5)	150	60
Snub cube	{3.3.3.3.4}	38 = 32(3)+6(4)	60	24

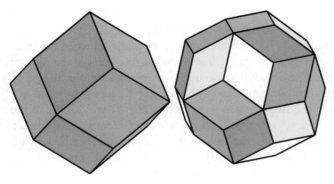

The rhombic dodecahedron with 12 faces and the rhombic triacontahedron with 30 faces are the duals of the cuboctahedron and icosidodecahedron.

which it will be recalled are more regular than the other Archimedean figures because all their edges are the same. Both these duals have faces which are rhombic (a rhombus is a parallelogram with all sides equal) and are known as the rhombic dodecahedron which has 12 faces and the rhombic triacontahedron which has 30 faces. They are shown above. The reader may feel confused in looking at them. The problem seems to be that the human eye is conditioned to assume that a rhombus is actually a square at an angle and so the brain attempts to interpret the image as bits of cubes stuck together.

The rhombic dodecahedron is important in crystallography and we shall meet it again in the next lecture when we discuss honeycombs. Its dihedral angle is exactly 120° as we shall see in a moment.

The rhombic triacontahedron in a sense has already been encountered since it is the figure common to the five cubes in the amazing compound of five cubes. If we look back at the diagram on page 42 showing the lines of intersection of the four coloured cubes with a face of the white cube we will see a rhombus at the centre of the white square. This rhombus is a face of the figure common to the five cubes and since each cube has six faces this makes the total 30. The triacontahedron above has been coloured with the faces corresponding to those of the five cubes.

A strange feature of these figures is that a zigzag belt of rhombi can be traced around the equator in various ways. In the case of the rhombic dodecahedron, the

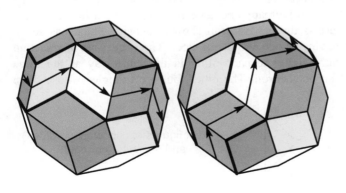

Two of the zigzag belts of the triacontahedron are highlighted.

belt has six rhombi and can be traced in four ways. These belts are the dual of the four equatorial hexagons of the cuboctahedron. Since the belt has six rhombi it follows that the dihedral angle is 120°, the same as the angle between the sides of a hexagon. In the case of the triacontahedron there are five belts of ten rhombi from which it follows that the dihedral angle of the triacontahedron is 144°.

A curious transformation we can perform with these figures is to remove a zigzag belt and then join the two remaining parts together. If we do this to the rhombic dodecahedron then we obtain a figure with just six rhombic faces which looks rather like a squashed cube. In the case of the rhombic triacontahedron we obtain an equally peculiar figure with twenty faces. (I have made no attempt to illustrate these because looking at them can drive the reader mad!)

These squashed figures can be called the rhombic hexahedron and rhombic icosahedron respectively. So we have discovered figures with 6, 12, 20 and 30 rhombic faces. These are examples of a general rule that we can construct a figure with $n(n-1)$ rhombic faces. These are sometimes called zonohedra.

Note that the rhombic faces of the dodecahedron and triacontahedron do not have the same shape. The ratio of the diagonals in the case of the dodecahedron is $\sqrt{2}$ whereas for the triacontahedron it is τ, the golden number.

Non-convex polyhedra

WE HAVE ALREADY met the four Kepler-Poinsot regular polyhedra which have pentagrams as faces or vertex figures. These are the only non-convex polyhedra whose faces are all the same.

However, if we allow any combination of polygons or polygrams as faces provided that the same arrangement occurs at each vertex (as for the Archimedean figures) and also allow the faces to intersect then other possibilities arise. There are in fact 53 such figures. We will not consider them in detail because it would take too much space but just mention a few in order to illustrate the variety of these figures.

An interesting new feature is that the faces that arise also include the octagram {8/3} and the decagram {10/3} as well as the pentagram and the various polygons that we have already encountered as faces of earlier figures.

An intriguing figure is the heptahedron shown opposite which has seven faces – it is unique in having an odd number of faces. Three faces are intersecting squares arranged at right angles to each other in the same way as the three equatorial squares of an octahedron. The other four are triangles whose three

Vertex figures of cuboctahedron (left) and heptahedron (right).

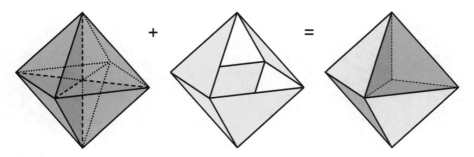

Three squares plus four triangles make a heptahedron {3.4,3.4}.

sides join the sides of the three squares. Each of the eight vertices is surrounded by two squares and two triangles. That is the same arrangement as in the cuboctahedron but, in the case of the heptahedron, the vertex figure crosses over itself as shown opposite. We can represent this by using a comma rather than a dot in the list of faces at the point where the sides of the vertex figure change direction thus {3.4, 3.4}. The heptahedron is a weird figure with strictly speaking only one side. It can be looked upon as an octahedron with alternate sections missing. We will meet the heptahedron again in the lecture on Topology.

Another relatively simple pair of figures can be derived from the cuboctahedron {3.4.3.4}. First take four intersecting hexagons arranged as the equatorial hexagons of the cuboctahedron. These can then be joined by the eight triangles of the cuboctahedron or by the six squares of the cuboctahedron as shown below. The resulting figures are known as the octahemioctahedron and cubohemioctahedron respectively. Each can be considered to be part of a cuboctahedron and together they make a complete cuboctahedron. One is one-sided and the other is two-sided.

A similar pair of figures can be obtained from the icosidodecahedron {3.5.3.5}. In this case there are six intersecting equatorial decagons and these can be joined by 20 triangles or by 12 pentagons.

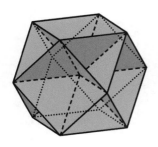

Four equatorial hexagons of a cuboctahedron.

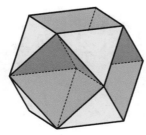

Octahemioctahedron {3.6,3.6}

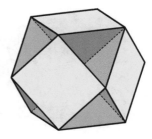

Cubohemioctahedron {4.6,4.6}

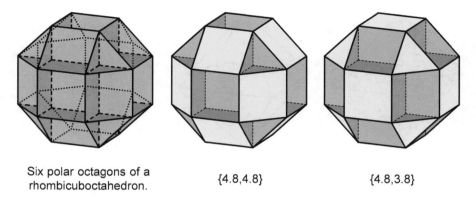

Six polar octagons of a rhombicuboctahedron.   {4.8,4.8}   {4.8,3.8}

Another pair of figures can be obtained from the rhombicuboctahedron {3.4.4.4} as shown above. This does not have equatorial polygons but instead has pairs of octagons on either side of the equatorial belts of squares and so positioned a bit like the arctic and antarctic circles of the earth. Moreover, there are three such belts of squares intersecting at right angles. So we start with six octagons in three parallel pairs intersecting each other at right angles. We can then join these either with twelve squares or with eight triangles and six squares.

A similar pair of figures can be obtained from the rhombicosidodecahedron {3.4.5.4}. In this case the framework is provided by 12 decagons arranged as six intersecting parallel pairs. The 12 decagons can then be joined by 30 squares or by 20 triangles and 12 pentagons.

Two figures involve six octagrams {8/3} arranged as the faces of a cube with pairs of vertices touching. The edges can then be joined in two ways as shown below. In one figure, six squares are arranged in pairs parallel to the octagrams and thus oriented as the faces of a cube; the remaining edges are then joined by eight triangles oriented as the faces of a dual octahedron. In the other figure the edges are joined by twelve squares arranged as six parallel pairs with two pairs perpendicular to each pair of octagrams; thus two parallel mid red squares and two parallel pale red squares join the red octagrams and so on.

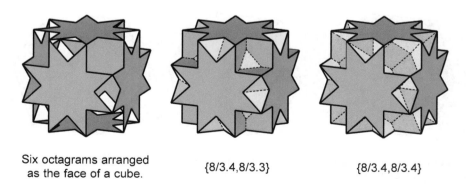

Six octagrams arranged as the face of a cube.   {8/3.4,8/3.3}   {8/3.4,8/3.4}

Two views of a model of the great dodecahemidodecahedron {10/3.5/2,10/3.5/2}.

Two figures involve decagrams {10/3} in much the same way. They have 12 decagrams arranged as the faces of a dodecahedron with pairs of vertices touching. Their sides are then joined by 20 triangles and 12 pentagons or by 20 hexagons.

Four figures involve 12 pentagrams arranged as in the faces of a dodecahedron with their vertices touching. In two cases the vertices of three adjacent pentagrams touch and the edges are then joined by 20 triangles or by 12 pentagons. In the other two cases the pentagrams are oriented so that they meet in pairs at the vertices and the edges are then joined by 12 pentagons or by 10 hexagons.

That very briefly outlines just 17 of this amazing group of 53 figures. Many of the others are very intricate and defy an outline description. But I must mention one more since it is the only one composed only of polygrams.

The so-called great dodecahemidodecahedron consists of twelve pentagrams interlinked with six decagrams. The six decagrams are arranged equatorially as if parallel to the pairs of faces of a dodecahedron. The pentagrams are in pairs parallel to and each side of a decagram. Each decagram meets each of the five others at two opposite vertices. The vertices are also joined by the pentagrams. It is shown above as a stereo pair.

Pentagonal tilings

As noted earlier, it is not possible to tile a plane with regular pentagons because the angle of a pentagon is 108° which is not a factor of 360°. But it is quite easy to tile a plane with pentagons that are not regular as shown overleaf.

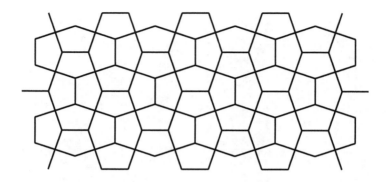

A plane can be tiled with equal but not regular pentagons.

However, quite recently (in mathematical terms) it was shown by Roger Penrose that it is possible to tile the plane with a mixture of regular pentagons with the occasional pentagram, rhombus and a boat-like shape which is part of a pentagram as shown below. There are also rules regarding which edges can be placed together. Given these rules it can be shown that the tiling never repeats exactly although portions will be found which match as closely as we like; such a tiling is said to be aperiodic. A curious feature is that if we attempt to assemble such a tiling then we will quite often get into an impossible position and have to

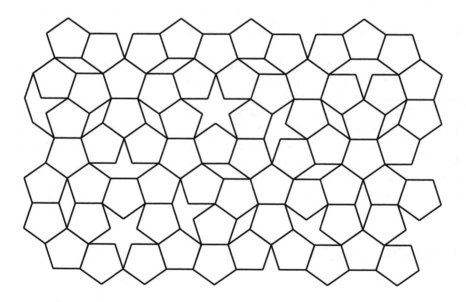

An aperiodic tiling with a pentagonal flavour.

backtrack. Moreover, the tiling almost shows fivefold symmetry since the rows of pentagons can be traced in five directions. Note also the various rings of ten pentagons.

In crystallography it has long been a golden rule that symmetry can be threefold, fourfold or sixfold but never fivefold. But here we see the strange possibility of crystals based on fivefold symmetry and indeed quasicrystals of alloys exhibiting such structures have been observed. Being in three dimensions these crystals have an icosahedral symmetry.

This is typical of much modern mathematics and verges on the ideas of chaos which we will discuss later. I suppose the trouble started with Gödel who loosely speaking showed that no fixed set of mathematical procedures can ever suffice to establish all mathematical truths; this showed that mathematics is not a completely cut and dried discipline.

Further reading

THE AMAZING BOOK *Flatland* by Abbott is a clear starting point for this lecture. *Mathematical Models* by Cundy and Rollett is a classic book and describes the regular tilings and gives details of the regular and Archimedean figures and guidance on their construction. *Polyhedron Models* by Wenninger covers all the regular figures including the 53 non-convex ones; it includes detailed instructions on how to make models and includes photographs of every one. Strangely, although the compound figure of five cubes is mentioned, constructional details are omitted although it gives details of the other compound figures. The five cubes are however described by Cundy and Rollett.

Pretty photographs of some figures will be found in *Mathematical Snapshots* by Steinhaus. The weird rhombic zonohedra are described in *Regular Polytopes* by Coxeter.

The pentagonal tilings and other aperiodic tilings are described in *The Emporer's New Mind* by Roger Penrose; this also includes a discussion of quasicrystals. An exhaustive discussion of tilings in general will be found in *Tilings and Patterns* by Grünbaum and Shepherd; this also includes the Penrose pentagonal and other tilings and describes in detail the various rules for matching the edges.

A rather different book is *Islamic Patterns* by Critchlow which contains many diagrams as well as an interesting discussion covering all the regular and semi-regular tilings.

The lithographs by Escher will be found in *Escher, The Complete Graphic Work* edited by Locher. Most will also be found in *Gödel, Escher, Bach* by Hofstadter. The paintings by Dali are illustrated in *Dali* by de Liaño.

See also Appendix A for an amusing relationship between the golden number τ and the square root of 2 and Appendix B for stereo images of various figures.

Exercises

1. How many different colours are required to colour a) the three regular tilings, b) the eight semi-regular tilings? Remember that adjacent tiles must be of a different colour but it doesn't matter if tiles of the same colour meet at a corner only. Note that a chessboard requires only two colours – now think about the others. (Only one tiling requires the maximum of 4 colours.)

2. A tetrahedron can be coloured in two different ways with four colours, one on each face. How many different ways can a cube be coloured with six colours, one on each face?

3.** The line joining the centres of adjacent hexagons in the tiling {3.4.6.4} is at right angles to the sides of the hexagons that it crosses. However, in the tiling {3.3.3.3.6} the line joining the centres of adjacent hexagons crosses the sides obliquely because the hexagons are rotated. What is the angle of rotation? Similarly what is the angle of rotation of the squares in the snub cube {3.3.3.3.4} compared with the corresponding squares in {3.4.4.4}?

3 The Fourth Dimension

THE PREVIOUS LECTURE described the regular figures in two and three dimensions. We now consider the regular figures in four dimensions and how we can get a glimpse of them by analogy with how a person in Flatland might get some appreciation of figures in three dimensions.

What is the fourth dimension?

PHILOSOPHERS, mathematicians and artists have long mused over the possibility of a fourth dimension. Einstein's theories of relativity were a big stimulus to (probably superficial) discussion of the fourth dimension at dinner parties in the early years of the twentieth century. Indeed, that seems to have been a time of increasing interest in the occult, meditation, spiritualism and similar topics. The contemplation of the fourth dimension was claimed to be good for gaining an expansion of consciousness, deeper insight into one's soul and so on. I would have thought that too much contemplation of the fourth dimension was more likely to result in a nasty headache!

There was some confusion regarding whether time is the fourth dimension. In Einstein's theories of relativity, time is indeed the fourth dimension but interwoven with the spatial dimensions as we shall discuss in Lecture 9. But in this lecture we shall be considering a fourth dimension in space. This is of course merely a mathematical figment of the imagination but even though we do not live in a world of four spatial dimensions and thus can have no direct experience of it nevertheless the properties of such a space and the possible regular figures can be predicted and described.

It is pretty hard to imagine a fourth spatial dimension but by analogy with how the Flatlanders could get some feel for the third dimension there are a number of approaches we can take. One way is to consider the possible sections of the various regular hypersolids. In other words what we might observe if a four-dimensional figure were to pass through our Spaceland in the same way that the Flatlanders could see the various sections of a cube even though they could not appreciate the cube as a whole. Another way is to imagine the projections of the objects onto our space. In this lecture we will concentrate on looking at sections whereas some projections as stereo images are shown in Appendix D.

Before looking at specific regular figures, it is worth recapping a few rules about three dimensions and stating the corresponding rules in four dimensions. And again we can use the Flatland analogy to help us.

For example, two planes intersect in a line in three dimensions (unless they are parallel). In four dimensions however, two planes generally meet only in a single point. If this seems strange then consider the analogy with lines in Flatland. Two lines in Flatland always meet in a point (unless parallel) but in our three-dimensional space two lines in general do not meet at all (unless by chance they are both in the same plane). A plane and a line generally meet in a point in our space but do not meet at all in four dimensions (unless by chance they are both in the same three-dimensional space).

The general rule is simple. In space of n dimensions, objects of p and q dimensions meet in an object of $p+q-n$ dimensions. Thus in $n=3$ dimensions, two planes ($p=q=2$ dimensions) meet in a line (2+2−3=1 dimension). If $p+q-n$ is zero then the objects meet in a point; if it is negative then they do not meet. So in four dimensions, two planes meet in a point (2+2−4=0) whereas a line and a plane do not in general meet at all.

Honeycombs

As a preliminary to delving into the magic of four dimensions, we will consider honeycombs in three dimensions. Although we normally associate the word honeycomb with the structure made by bees, it is used technically to mean the subdivision of a space into many equal parts. In two dimensions, this becomes the tilings which we saw in the previous lecture. Moreover, the tilings proved to be rather like some of the regular three-dimensional figures but opened flat. So considering honeycombs in three dimensions is a natural prelude to investigating four dimensions.

Perhaps surprisingly, the only regular honeycomb is that made of cubes. The Schläfli symbol for the cubic honeycomb is {4, 3, 4}. We can consider this as an overlapping combination of {4, 3} and {3, 4}. The {4, 3} represents the cube and the {3, 4} represents the vertex figure formed where eight cubes meet and is of

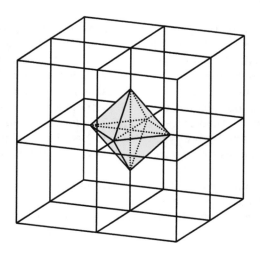

The vertex figure of the honeycomb of cubes is an octahedron.

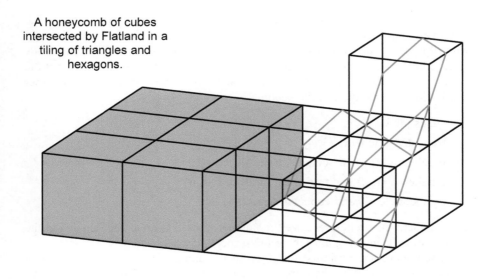

A honeycomb of cubes intersected by Flatland in a tiling of triangles and hexagons.

course the octahedron. Note that six edges meet at each vertex corresponding to the fact that the octahedral vertex figure has six vertices.

Remember that the vertex figure is that obtained by joining the midpoints of the edges which meet at the vertex concerned. By going up a dimension, the vertex figure becomes a polyhedron rather than a polygon.

It is interesting to consider what the Flatlanders would see if a piece of honeycomb of cubes were to pass through Flatland. If it were to pass face first then it would appear suddenly as a section of square tiling {4, 4}. On the other hand if the honeycomb were at an angle so that the individual cubes were approaching vertex first then the section would be a mixture of triangles and hexagons. And indeed when some of the cubes are exactly halfway then they appear as regular hexagons with the result that the pattern of the cross-section is the tiling {3.6.3.6}. Various cross-sections are shown above and below.

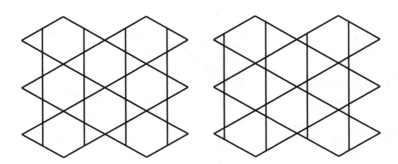

Two cross-sections through a honeycomb of cubes.

That space can be subdivided into cubes is obvious. We might think intuitively that space could also be subdivided into lots of tetrahedra by analogy with tiling the plane with triangles. But this is not so. However, if we allow the cells to be Archimedean polyhedra as well and allow mixtures but insist that the same number surround each edge and each vertex then there are four other possibilities so that the full list becomes

> cubes (4, 8)
>
> truncated octahedra (3, 4)
>
> tetrahedra (2, 8) and octahedra (2, 6)
>
> tetrahedra (1, 2) and truncated tetrahedra (3, 6)
>
> octahedra (1, 2) and cuboctahedra (2, 4)

The numbers in parentheses give the number of cells around each edge and the number of cells meeting at each vertex respectively.

The honeycomb of a mixture of tetrahedra and octahedra shown below in a sense is the analogue of the tiling of triangles in the plane. The tetrahedra have two different orientations and four of each orientation meet at each vertex. Two octahedra and one of each orientation of tetrahedra meet at each edge. Altogether no less than fourteen cells and twelve edges meet at each vertex. Note how this honeycomb is formed by a series of parallel planes at four different orientations. These planes containing the faces of the cells exhibit the regular tiling of triangles. This honeycomb appears in the lithograph by Escher entitled *Flatworms*.

Then there is the honeycomb of tetrahedra and truncated tetrahedra (opposite and above). Both occur in two different orientations. Two truncated tetrahedra of one orientation, one of the other and a tetrahedron meet at each edge. Eight cells and six edges meet at each vertex. Like the honeycomb of tetrahedra and octahedra it is formed by cutting space by a series of parallel planes at four different orientations; in the former case four planes go through each vertex, but

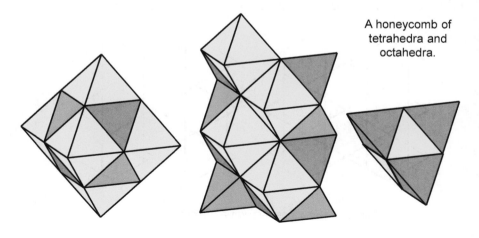

A honeycomb of tetrahedra and octahedra.

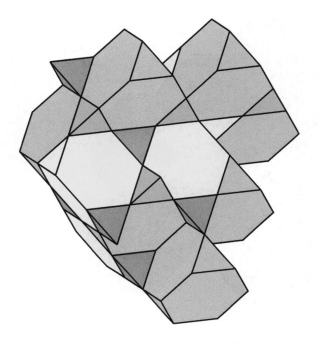

A honeycomb of tetrahedra and truncated tetrahedra.

in this case just three planes. The faces of the figures on these planes show the familiar tiling {3.6.3.6}.

The honeycomb of octahedra and cuboctahedra shown below can be obtained by taking the cubic honeycomb and making a hole at each vertex. Since the vertex figure is an octahedron, it follows that the holes will be octahedra. If we make them the right size then the cubes become reduced to cuboctahedra. Six cells and eight edges meet at each vertex.

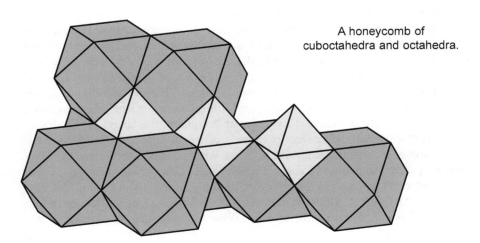

A honeycomb of cuboctahedra and octahedra.

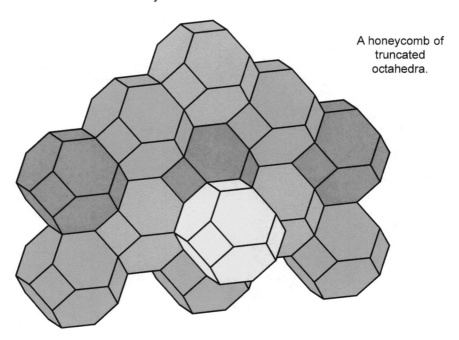

A honeycomb of truncated octahedra.

Finally, the truncated octahedra perhaps make the prettiest honeycomb; it has a certain elegance because it has the least number of polyhedra meeting at each edge and vertex. We will meet it again when we discuss soap bubbles in Lecture 6. The arrangement of the truncated octahedra within this honeycomb is rather harder to understand than for the other honeycombs. One way is to observe that the polyhedra can be seen as forming layers and within each layer they are arranged as for the tiling of hexagons (one layer in the diagram above has green, blue and red polyhedra arranged so that no two of the same colour touch). Another layer can then be placed on top and one such polyhedron is shown coloured yellow. Note that it touches polyhedra of all colours in the layer below.

The reader might like to consider what other tilings are exhibited by sections of the various honeycombs. We have already seen that sections exhibit the regular {4, 4} and {3, 6} and the mixed {3.6.3.6}. Some others occur as well.

Other less regular honeycombs are possible. We can make a honeycomb of triangular prisms and one of hexagonal prisms. These naturally arise by considering the triangular and hexagonal tilings of the plane and imagining similar parallel planes whose distance apart is equal to the length of the side of the tiles. But they are somewhat irregular because all the edges do not have the same number of prisms around them. If we do the same with the square tiling then we get the cubic honeycomb again because a square prism is simply a cube.

Another honeycomb is that made of rhombic dodecahedra (one of the dual Archimedean figures). Although three meet at each edge, some vertices have four and some have six around them so that it is not really regular. Nevertheless, it is an important figure in crystallography. We will meet it again later in this lecture.

The 4-simplex

THE TETRAHEDRON is the simplest figure in three dimensions and is obtained by taking a triangle which is the simplest figure in two dimensions and adding a fourth point in the extra dimension equidistant from the vertices of the triangle. Similarly, starting with a tetrahedron we add a fifth point in the new dimension equidistant from the other four. The result is the regular 4-simplex which has five vertices, ten edges, ten triangular faces and five tetrahedral cells. The five tetrahedra form the "surface" of the 4-simplex.

The diagram below shows a triangle ABC, a tetrahedron $ABCD$ and a 4-simplex $ABCDE$. The simplex has five tetrahedral cells and these are obtained by taking any four of the vertices. This can be done in five ways (the omitted vertex can be chosen in five ways obviously). Similarly, it has ten edges because there is an edge corresponding to the ten ways in which two vertices can be chosen from five. And it has ten triangular faces because three vertices can also be chosen in ten ways (each choice of three corresponds to choosing the remaining two). Consequently, there is a face opposite each edge where the face is identified by the three vertices not being on the edge; thus the face ABC is opposite the edge DE. Moreover, each cell is opposite a vertex; thus the cell $ABCD$ is opposite the vertex E. Note the similarity to the tetrahedron in which each edge has an opposite edge and each vertex has an opposite face.

Another interesting feature is that each edge belongs to three faces. Thus AB is an edge of ABC, ABD and ABE. In a three-dimensional figure each edge only belongs to two faces and in four dimensions this corresponds to the fact that each face belongs to two cells. Thus the face ABC belongs to the cells $ABCD$ and $ABCE$. Finally each vertex belongs to four cells. Thus the vertex A belongs to the cells $ACDE$, $ABDE$, $ABCE$ and $ABCD$.

The Schläfli symbol for the 4-simplex is $\{3, 3, 3\}$. The first two 3s show that the cells are tetrahedra $\{3, 3\}$. The fact that the last two numbers are 3s as well shows that the vertex figure is also a tetrahedron and indeed this can be confirmed by noting that if we join the four points in the middle of the four edges meeting at a vertex then we get the same arrangement of points and lines as in a tetrahedron.

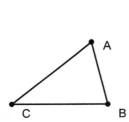 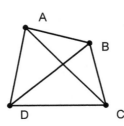 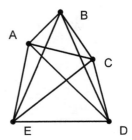

A triangle $\{3\}$, tetrahedron $\{3, 3\}$ and 4-simplex $\{3, 3, 3\}$.

We will now consider what we Spacelanders would see if a 4-simplex passed through our space. As in the case of three-dimensional objects passing through Flatland, the appearance depends upon the orientation.

If the 4-simplex approaches vertex first then it will first appear as a Point which will immediately become a growing Tetrahedron until the opposite cell (which is a tetrahedron) is in our space. And then the Tetrahedron will suddenly vanish as the 4-simplex moves away. If it approaches cell first then the reverse will happen – a Tetrahedron will suddenly appear and then slowly shrink to a Point. (Remember that the 4-simplex, like the tetrahedron, is not centrally symmetric since each vertex is opposite a cell rather than another vertex and each edge is opposite a face.)

On the other hand, suppose the 4-simplex approaches with edge DE first. It will first appear as a Line (in red below) and then this will turn into three lines close together connecting two small triangles at each end, in other words a Triangular Prism (in blue). That this should be the case can be seen by studying the first diagram which shows the position when the 4-simplex is a little way through our space.

It is important to realize that a plane in a general position in four dimensions will intersect our space in a single line and that a line will generally meet our space in a single point. This can be seen by considering the three-dimensional analogy whereby a line generally intersects a plane in a single point; the exceptions are when the line is wholly in the plane or is parallel to it in which case it doesn't meet it at all.

So the faces ADE, BDE, and CDE each meet our space in one of the long edges of the prism. The other faces except ABC meet our space in the six short lines of the two triangular ends of the prism. The face ABC doesn't meet our space at all because it is parallel to it.

As the 4-simplex progresses, the Prism changes, the triangular ends grow larger and closer together as in the second diagram until finally the triangles merge into one Triangle which happens as the face ABC opposite the initial edge DE reaches our space. The reverse happens of course if the simplex approaches face first. It starts as a Triangle, then becomes a Prism and ends up as a Line.

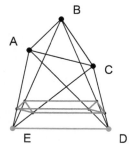 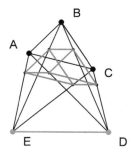

Two views of a 4-simplex intersecting our space in a triangular prism.

The hypercube and the 16-cell

WE WILL NOW consider two other simple four-dimensional figures which are analogous to the cube and its dual, the octahedron, in three dimensions.

The hypercube (or tesseract as it is often called) corresponds to the cube. A cube is formed by taking a square in two dimensions and connecting it to an equal square in a parallel plane. Similarly a hypercube is obtained by taking a cube and connecting it point by point to another cube in a "parallel" space. The hypercube has 16 vertices, 32 edges, 24 square faces and 8 cubic cells. Representations of a square, cube and hypercube are shown below.

The Schläfli symbol for a hypercube is $\{4, 3, 3\}$. The cells are cubes $\{4, 3\}$ and the vertex figure is a tetrahedron $\{3, 3\}$. Observe that three square faces meet at each edge. This is also encoded in the Schläfli symbol – the initial 4 says that the faces have four edges and the final 3 says that the edges are on three faces. Moreover, each vertex is part of four cells.

Each cubic cell of a hypercube is opposite another cubic cell in much the same way that each square face of a cube is opposite another square face.

The four-dimensional figure corresponding to the octahedron is the 16-cell. A convenient naming convention is simply to use the number of cells which is really what we do in three dimensions only the names are glamorized by being Greek. Remember that octahedron just means 8-faced. We can also refer to the hypercube as the 8-cell and the 4-simplex as the 5-cell. One way of constructing an octahedron is to take three lines (axes) at right angles and then to mark the vertices as the points on the lines which are unit distance from the centre. We construct the 16-cell in the same way by taking two points on each of four lines at right angles. The 16-cell has 8 vertices, 24 edges, 32 triangular faces and 16 tetrahedral cells. Representations of the octahedron and 16-cell are shown overleaf in which dotted lines represent the axes. Note that the two-dimensional analogue is also a square.

The Schläfli symbol for the 16-cell is $\{3, 3, 4\}$. This shows that the cells are tetrahedra $\{3, 3\}$ and the vertex figure is an octahedron $\{3, 4\}$. Four triangular faces meet at each edge and each vertex belongs to eight cells.

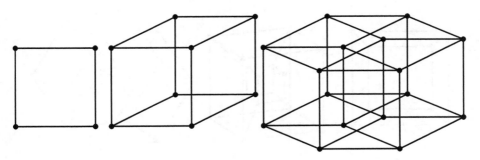

A square $\{4\}$, cube $\{4, 3\}$ and hypercube $\{4, 3, 3\}$.

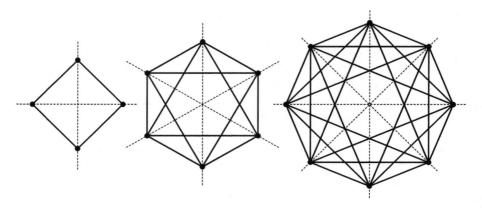

A square {4}, octahedron {3, 4} and 16-cell {3, 3, 4}.

Note that the hypercube and 16-cell are duals and that the 4-simplex is self-dual. In four dimensions, edges and faces are dual concepts as are vertices and cells.

Other representations of the hypercube and 16-cell are possible. The representation of the hypercube shown earlier was obtained by taking the engineering representation of the cube and then joining it to a copy. Another possible representation of a cube is one square within another which is the perspective view obtained by looking at it straight on; the far side looks smaller of course because it is further away. A similar representation of a hypercube is one cube within another as shown below. That the hypercube has eight cubes as cells is quite easily seen in this representation. Two of the cubes are the outer and inner one and the remaining six are the rather distorted shapes that have one face of the inner cube and one face of the outer cube as opposite faces.

However, this representation is not very helpful for our purposes because it is not an orthogonal view and as a consequence lines which are truly parallel do not appear to be parallel. So we need to use an orthogonal projection of a

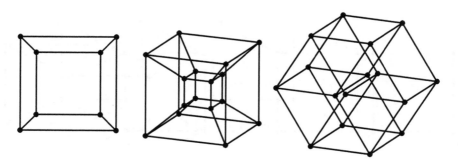

Other representations of a cube and hypercube.

 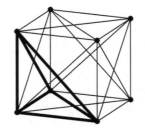 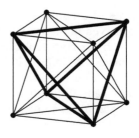

16-cells highlighting individual tetrahedral cells.

hypercube. Care is needed to ensure that lines and points do not coincide as for example happens with an orthogonal projection of a cube looking straight at the centre of a face – all we see is a single square! For the hypercube, we use a projection from near the centre of an edge. The result is the third diagram shown opposite. From exactly the centre of an edge would cause the two central lines to coincide hence the slight deviation. Note that this view is similar to the engineering representation shown earlier.

An interesting alternative representation of the 16-cell is a cube with the face diagonals drawn. This is in fact the orthogonal view obtained by looking at a 16-cell from the centre of one of its cells. Each face of the cube plus its diagonals is one of the tetrahedral cells of the 16-cell which appear flat in this view. Since the cube has six faces that accounts for six out of the sixteen cells of the 16-cell. Eight others are at each corner of the cube and comprise the corner plus the three points closest to it. The other two are simply the two tetrahedra embedded in the cube which formed the stella octangula compound figure discussed in Lecture 2; thus we take two corners of the top face plus the two corners of the bottom face which do not correspond and of course this can be done in two ways. Some of the cells are shown in bold in the diagrams above.

We will now consider what we might see when hypercubes and 16-cells pass through our space. There are four key possible orientations in both cases, vertex first, edge first, face first and cell first so quite a lot of different views are possible. (In the diagrams that follow the initial item is shown in red and the various subsequent cross-sections in blue.) We start with the hypercube.

If a Hypercube approaches cell first then it suddenly appears as a Cube and then suddenly disappears much as a Cube passing through Flatland face first is seen as a Square which suddenly appears and then disappears.

If a Hypercube approaches face first then the sequence is first a Square and then the four corners double up to give a very shallow square box formed by joining two parallel squares together (a form of cuboid). As the Hypercube progresses, the two squares move apart until the figure becomes a Cube. As it progresses further the original squares continue to move apart until the Hypercube is exactly halfway and eight of its vertices are in our space. The figure is then a Cuboid with square ends and whose long edges are $\sqrt{2}$ times the sides of the square. The process then reverses. This is a similar sequence to that observed in Flatland as a Cube passes through edge first.

If the Hypercube approaches vertex first then it will first appear as a Point which will immediately become a growing Tetrahedron (remember that the vertex figure is a tetrahedron). After a while the four nearest vertices will reach our space. When this happens each vertex of the tetrahedron divides into three because three further lines emanate from the vertex of the hypercube. The result is that the tetrahedron becomes truncated and eventually turns into a Truncated Tetrahedron (one of the Archimedean figures) with four regular hexagonal faces and four triangular faces. At this point the Hypercube is at a position where the four cubic cells containing the vertex which first entered our space are exactly halfway through and are sliced by our space in a hexagon – remember that the Flatlanders saw a hexagon when a cube passed through Flatland vertex first. The four cubic cells remote from the leading vertex of the Hypercube are only a little way into our space and so are sliced in a triangle which explains the triangular faces of the Truncated Tetrahedron. As the Hypercube progresses the Truncated Tetrahedron becomes further truncated until it eventually becomes an Octahedron. This happens when the Hypercube is exactly halfway and six of its vertices are in our space. The process then repeats in reverse. The Octahedron turns via a different Truncated Tetrahedron into a Tetrahedron which then shrinks to a Point. The diagrams opposite show the situation (1) when the tetrahedron is at a maximum, (2) when the truncated tetrahedron is regular, and (3) the midway octahedron.

Finally, if a Hypercube approaches edge first then the sequence is first a Line, then a Triangular Prism (much like the 4-simplex). The next change is when a group of six vertices reach our space. Each vertex of the prism then divides into two so that the ends of the prism become irregular hexagons. When the Hypercube is halfway these hexagons become regular so that we see a Hexagonal Prism. The process then reverses. Remember that each edge of a hypercube belongs to three different cubic cells and these cells are passing through our space edge first – this accounts for the rectangular faces. Furthermore, each vertex of the initial edge of the hypercube belongs to one other cubic cell because four cells meet at each vertex of a hypercube. These two other cubes are passing through our space vertex first and this explains the initial triangle which becomes a hexagon at the halfway position. The diagrams opposite show (1) when the triangular prism is half its maximum size, (2) when the triangular prism is at a maximum, and (3) the midway hexagonal prism.

These various sequences may seem rather complicated but the key to understanding what happens is to note that as the Hypercube moves along, groups of its vertices pass through our space (the actual number depends upon the orientation) and when this happens there is a significant change in the properties of the shape as seen in our space. The sequences are summarized in the table on page 76; the entries with numbers correspond to sections where that number of vertices are in our space. (Note that the numbers 88, 484, 2662, 14641 are rows of Pascal's triangle multiplied by an appropriate power of 2.)

3 The Fourth Dimension

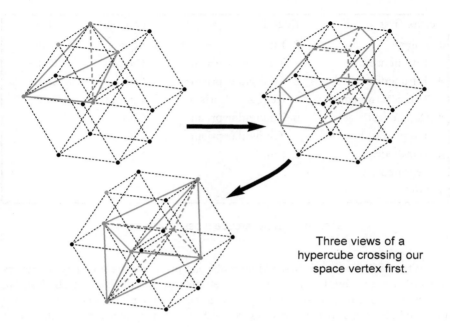

Three views of a hypercube crossing our space vertex first.

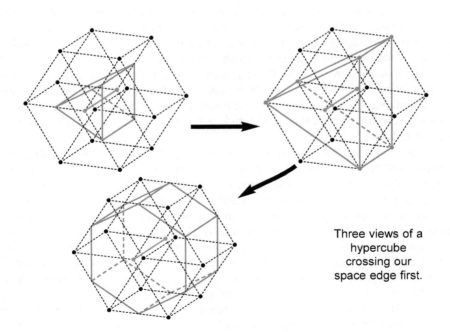

Three views of a hypercube crossing our space edge first.

Vertex first	Edge first	Face first	Cell first
1 Point	2 Line	4 Square	8 Cube
Tetrahedron	Triangular prism	Cuboid	Cube
4 Tetrahedron	6 Triangular prism	Cube	8 Cube
Trunc Tetrahedron	Hexagonal prism	8 Cuboid	
6 Octahedron	6 Triangular prism	Cube	
Trunc Tetrahedron	Triangular prism	Cuboid	
4 Tetrahedron	2 Line	4 Square	
Tetrahedron			
1 Point			

Sections of a hypercube crossing our space.

What we have seen is that a Hypercube can present quite a variety of figures to us as it passes by. It can appear as a tetrahedron, a truncated tetrahedron, an octahedron, a cube, a triangular prism, a hexagonal prism and also various intermediate forms. It is interesting to observe that our investigations show that these figures can be continuously transformed from one to another. Moreover, we have only considered the effect as the hypercube traverses our space with a fixed orientation – we have not considered how the figures change if the hypercube were to rotate.

We will now consider the dual 16-cell. In some ways the 16-cell is a bit easier because it has fewer vertices and edges. But on the other hand the hypercube is perhaps easier to understand intuitively because we can readily (perhaps) see the various cubes of which it is composed.

If a 16-cell approaches vertex first then the initial Point becomes a growing Octahedron which reaches a maximum when the 16-cell is halfway through our space and then shrinks again down to a Point. This is similar to the way in which an Octahedron passing through Flatland vertex first appears as a Square which grows and then contracts.

If a 16-cell approaches cell first then it immediately appears as a Tetrahedron because the cells are tetrahedra and then each vertex breaks into three points, the edges of the tetrahedron become flattened so that long rectangles appear between the adjacent faces of the tetrahedron and the vertices turn into small triangles. When the 16-cell is halfway, the rectangles become squares and the whole figure is then a Cuboctahedron with six square faces and eight triangular faces. The process then reverses with the newly created triangles eventually becoming the faces of the tetrahedral cell of the 16-cell opposite the cell which originally entered our space. The diagram opposite shows the flattened tetrahedron (when one-fifth of the way) and the midway cuboctahedron.

If a 16-cell approaches edge first then the sequence is first a Line, then a long Square Prism with square pyramid caps. As the 16-cell progresses, the

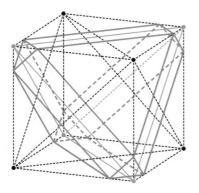 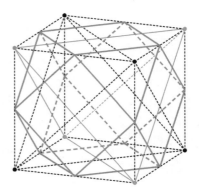

A 16-cell crossing our space cell first.

squares approach and when the 16-cell is halfway (and then four vertices of the 16-cell are in our space), the squares meet and the resulting figure becomes two square pyramids arranged base to base. Note that although it has eight triangular sides they are not equilateral and so the figure is not a regular octahedron because the angles are not the same. In fact the dihedral angle between the base and sides of the square pyramids is 45° whereas the corresponding angle in a regular octahedron is 54° 44'. The process then reverses. The diagram below shows two views of this sequence, the prism plus caps (when one-eighth of the way) and the midway double square pyramid.

Finally, if a 16-cell approaches face first then the sequence starts with a Triangle. The corners of the triangle then become four points and the figure consists of two parallel triangles joined along their edges by pairs of thin four-sided figures (trapezia) and at their ends by pairs of triangles; in other words it is a shallow Triangular Prism plus a border of trapezia and triangles. As the 16-cell progresses the two big triangles shrink until they vanish when the figure

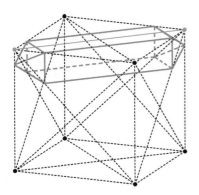 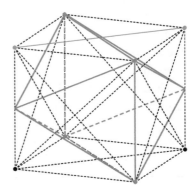

A 16-cell crossing our space edge first.

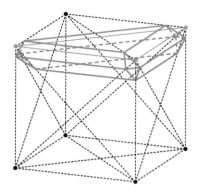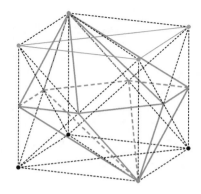

A 16-cell crossing our space face first.

becomes two hexagonal pyramids base to base – this occurs when the 16-cell is halfway and two vertices of the 16-cell are in our space and are in fact the apices of the pyramids. The process then reverses. This rather complex sequence can be explained by noting that each face of the 16-cell belongs to two tetrahedral cells which pass through our space face first and account for the big triangles that shrink. In addition, each edge of the original face belongs to two other tetrahedra and these pass through edge first (but at an angle) thus explaining the trapezia. Finally, each vertex of the original face belongs to two further tetrahedral cells and these pass through vertex first (again at an angle) and explain the small triangles. The diagram shows the triangular prism with border (when one-twelfth of the way) and the midway double hexagonal pyramid.

The various sequences are summarized in the table below in a similar way to the sequences for the hypercube.

Vertex first	Edge first	Face first	Cell first
1 Point	2 Line	3 Triangle	4 Tetrahedron
Octahedron	Square prism + pyramid caps	Triangular prism + border	Cuboctahedron
6 Octahedron	4 Double square pyramid	2 Double hex pyramid	4 Tetrahedron
Octahedron	Square prism + pyramid caps	Triangular prism + border	
1 Point	2 Line	3 Triangle	

Sections of a 16-cell crossing our space.

Other regular convex figures

THERE ARE THREE other regular hypersolids making a total of six compared with the five Platonic solids in three dimensions. See the table below.

The first new figure is the 24-cell which is related to both the Hypercube and the dual 16-cell. The 24-cell has 24 octahedral cells, 24 vertices, 96 triangular faces and 96 edges. Its Schläfli symbol is {3, 4, 3} from which we see that the vertex figure is a cube. Three triangular faces meet on each edge and eight cells meet at each vertex. The 24-cell has the interesting property of being self-dual; it has no analogue in three dimensions.

The other two figures are the 120-cell and 600-cell which form a dual pair as is clear from their Schläfli symbols being the reverse of each other. They are analogous to the dodecahedron and icosahedron in three dimensions. Clearly these are complex figures and virtually impossible for humble three-dimensional beings to comprehend. I have no doubt that the 120-cell is very beautiful to a hyperperson being composed of 120 dodecahedral cells with a total of 720 pentagonal faces. Probably the 600-cell composed of 600 tetrahedral cells with a total of 1200 triangular faces is just rather too much!

Observe that the formula $C-F+E-V$ always has value zero. This is the four-dimensional analogue of Euler's formula $F-E+V$ in three dimensions which as we saw in the previous lecture always has the value 2.

The item labelled $C//V$ gives the number of cells meeting at each vertex. This is computed by multiplying the number of vertices in each cell by the number of cells and then dividing by the number of vertices in the overall figure. Thus in the case of the hypercube, the cells are cubes and have 8 vertices each so the calculation is $8 \times C / V = 8 \times 8 / 16 = 4$. In the case of the 600-cell no less than 20 tetrahedral cells meet at each vertex which is a scary indication of its richness.

The item labelled $F//E$ gives the number of faces meeting at each edge. This can be similarly computed by multiplying the number of edges of each face by the number of faces and then dividing by the overall number of edges. In fact the result is always the last number of the Schläfli symbol.

We will not explore the three new figures in detail as we did the 4-simplex, hypercube and 16-cell. Suffice it to say that sections of the 24-cell include cubes, octahedra and cuboctahedra and sections of the 120-cell and 600-cell include tetrahedra, dodecahedra, icosahedra and icosidodecahedra.

Object	Symbol	Cells	Faces	Edges	Vertices	$C//V$	$F//E$
4-simplex	{3, 3, 3}	5	10	10	5	4	3
Hypercube	{4, 3, 3}	8	24	32	16	4	3
16-cell	{3, 3, 4}	16	32	24	8	8	4
24-cell	{3, 4, 3}	24	96	96	24	8	3
120-cell	{5, 3, 3}	120	720	1200	600	4	3
600-cell	{3, 3, 5}	600	1200	720	120	20	5

Non-convex regular figures

THE READER will recall from the last lecture that there are four regular non-convex polyhedra which are usually known as the Kepler–Poinsot polyhedra. They have pentagrams as faces or as vertex figures. The question arises as to whether such amazing figures exist in four dimensions. The answer is Yes. There are in fact ten of them and their basic statistics are listed in the table.

As expected they all have pentagrams somewhere in their Schläfli symbols. They are closely related to the 120-cell and 600-cell in the same way that the Kepler–Poinsot polyhedra are closely related to the dodecahedron and icosahedron. Two of them are self-dual and the other eight form four dual pairs.

All the Kepler–Poinsot polyhedra occur as cells of one or more of these four-dimensional figures. I am sure that some have great beauty. My favourite Kepler–Poinsot polyhedron is the Great Dodecahedron {5, 5/2} and this occurs as the cells of the self-dual {5, 5/2, 5} whose vertex figure is the dual Small Stellated Dodecahedron and as the cells of {5, 5/2, 3} whose vertex figure is the Great Stellated Dodecahedron.

The density of these figures is odd. The reader may recall that the density of a polyhedron is the number of times a ray from the centre to the outside crosses a face (counting crossing the core of a pentagram as two). Similarly, the density of a four-dimensional figure is the number of cells that a ray crosses. Curiously, the density of these figures reveals an extraordinary sequence of numbers.

The more attractive Kepler–Poinsot polygons are those with the lower density. By analogy perhaps the same applies to these four-dimensional figures. If this is so then those with density 191 must surely be nasty convoluted beasts whereas those with density 4 and 6 will be the most attractive. I am sure that the self-dual {5, 5/2, 5} is gorgeous.

We will not contemplate the appearance of these figures as they cross our space. But clearly they will present an amazing galaxy of forms based on dodecahedra, icosahedra and the Kepler–Poinsot polyhedra as well as the simpler

Symbol	Cells	Faces	Edges	Vertices	C//V	F//E	Density
{5/2, 5, 3}	120	720	1200	120	12	3	4
{3, 5, 5/2}	120	1200	720	120	12	5	4
{5, 5/2, 5}	120	720	720	120	12	5	6
{5/2, 3, 5}	120	720	720	120	20	5	20
{5, 3, 5/2}	120	720	720	120	20	5	20
{5/2, 5, 5/2}	120	720	720	120	12	5	66
{3, 5/2, 5}	120	1200	720	120	12	5	76
{5, 5/2, 3}	120	7200	1200	120	12	3	76
{5/2, 3, 3}	120	720	1200	600	20	3	191
{3, 3, 5/2}	600	1200	720	120	20	5	191

polyhedra. Moreover, just as some cross-sections of the Kepler–Poinsot polyhedra by Flatland will appear as several separate pieces, many sections of these non-convex four-dimensional figures by our space will comprise separate pieces.

Honeycombs, five dimensions and more

WE WILL NOT consider four-dimensional figures akin to the Archimedean figures in which the cells are not all the same in this lecture. The reader will not be surprised to learn that such figures do exist but it would be more likely to bring on that headache rather than improve the repose of our souls to contemplate such complexity. But see Appendix C.

However, we will say a few words about honeycombs. We started this lecture by noting that the only really regular honeycomb in three dimensions is that formed of cubes {4, 3, 4}. There are in fact three regular honeycombs in four dimensions with symbols {4, 3, 3, 4}, {3, 3, 4, 3}, and {3, 4, 3, 3}.

The first shows that fourspace can be filled with hypercubes {4, 3, 3} as expected. The other two are duals; one shows that fourspace can be filled with 16-cells {3, 3, 4} and the other shows that fourspace can be filled with 24-cells {3, 4, 3} which is perhaps rather surprising and reveals just one more amazing aspect of four-dimensional space.

Indeed, the consideration of honeycombs is perhaps the best way of introducing the 24-cell. The cubic honeycomb in two dimensions is of course the tiling of squares. We can divide a square into four equal parts each consisting of a triangle whose base is one of the sides of the square and such that the apices of the four triangles are the centre of the square. Suppose now that we take a tiling of squares and do this subdivision to alternate squares and then add each triangle to the adjacent square that has not been subdivided. Then clearly the squares that were not subdivided plus the triangles added to them must also produce a tiling. However, it simply turns out to be another square tiling with larger squares and rotated through 45° as shown below.

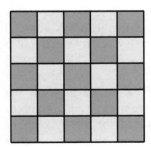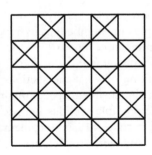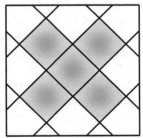

Dividing alternate squares in the tiling into four triangles and then adding them to the remaining adjacent squares gives another square tiling but at 45°.

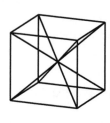 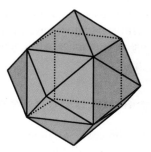 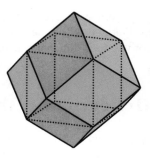

| A cube divided into six square pyramids. | A cube with a square pyramid on each face. | A rhombic dodecahedron circumscribing a cube. |

We can do the same thing with the honeycomb of cubes in three dimensions. A cube can be subdivided into six square pyramids; the base of each pyramid is a face of the cube and the apices of the pyramids meet at the centre of the cube as shown in the first diagram above. Suppose we now take a cubic honeycomb and do this subdivision to alternate cubes and then add each pyramid to the adjacent cube that has not been subdivided.

Clearly we now have a new honeycomb dividing three-dimensional space and in fact this is the honeycomb of rhombic dodecahedra briefly mentioned earlier. Each cube plus six pyramids forms a rhombic dodecahedron as shown in the second and third diagrams above. The resulting honeycomb is shown opposite. A layer of green and red polyhedra lies in one plane and a layer of yellow and blue ones forms an adjacent plane. It is not that regular since some vertices have four and some six cells around them.

We can obviously do the same thing with the four-dimensional honeycomb of hypercubes. A hypercube has eight cubic cells and so can be subdivided into eight hyperpyramids each with a cubic cell as its base and all meeting at the centre of the hypercube. The six side cells of these hyperpyramids are (of course) square pyramids. If we now do this subdivision to alternate hypercubes of the honeycomb and add the hyperpyramids to the adjacent undivided hypercubes then again we must get a new regular subdivision of four-dimensional space. It turns out that a hypercube plus the eight adjacent hyperpyramids forms a 24-cell $\{3, 4, 3\}$. And so we obtain the honeycomb $\{3, 4, 3, 3\}$. By duality there is also the honeycomb of 16-cells with symbol $\{3, 3, 4, 3\}$.

It is worth a brief mention of what the sections of these four-dimensional honeycombs by our three-dimensional space might be. In the case of the hypercubic honeycomb, the possible sections include the normal honeycomb of cubes, the honeycomb of tetrahedra and octahedra and the honeycomb of tetrahedra and truncated tetrahedra (remember that possible sections of a hypercube vertex first are tetrahedra, truncated tetrahedra and octahedra). Sections of the honeycombs of 16-cells and 24-cells include the honeycomb of cuboctahedra and octahedra.

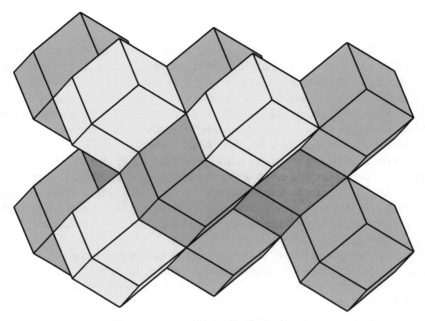

A honeycomb of rhombic dodecahedra.

We have seen that four dimensions reveals more complexity than three having six convex regular figures as opposed to five and ten non-convex ones as opposed to four. It also has three regular honeycombs as opposed to just one. We might expect things to get yet more elaborate in five dimensions. But it doesn't. All that exist in five (and more) dimensions are the analogues of the tetrahedron, cube and octahedron and the honeycombs of supercubes.

So four dimensions is uniquely rich and the self-dual 24-cell is itself unique in having no analogue in three dimensions.

These various statistics are summarized in the table below.

Dimensions Objects	Two	Three	Four	Five +
Regular convex	infinite triangle, square, pentagon, etc	five tetra-, hexa- octa-, dodeca- icosahedron	six 5-cell, 8-cell 16-cell, 24-cell 120-cell, 600-cell	three n-simplex supercube its dual
Regular nonconvex	infinite pentagram, etc	four Kepler–Poinsot	ten {5/2, 5, 3} etc	none –
Honeycombs	three triangle, square, hexagon tilings	one cubic	three hypercubic 16-cell, 24-cell	one supercubic

Nets

ANOTHER WAY of describing a figure in three dimensions is to give the so-called net from which it can be constructed by folding. Thus the net of a cube is a figure made of six squares arranged as shown below. To complete the description we need to say which sides are to be considered glued together and strictly in which direction although this is usually obvious. The diagram shows arrows and letters against some of the sides; the others are obvious.

We can now explain to a Flatlander that a cube is the figure obtained by rotating the faces along the edges until the marked edges join up. Of course he won't understand how one could rotate something about an edge because to him rotation only occurs about points. Nevertheless, we could explain that if it were folded up and somehow he could negotiate the corners then he would find that going out of the far right hand square by the edge "a" will miraculously bring him into the far left square by its edge "a".

We can now extrapolate this to the representation of a hypercube in three dimensions. The required net comprises eight cubes, four of which are in a row and the other four are around the second one of the row. The diagrams opposite hopefully illustrate this. The one on the left shows all the edges whereas that on the right has the hidden edges removed.

Now all we have to do is to fold it up. We need to rotate the cubes about the appropriate faces. But of course we Spacelanders only know about rotation about lines and cannot comprehend how we might rotate something about a plane. Oh well, at least we can do the identifications and so can understand that if we were able to fold the thing up to form a hypercube, then the cube on the right would be joined to the cube on the left and so on. Some of the faces are lettered to show how they are identified. As a result, if we were able to traverse the cells of the hypercube then by going out of the far side of the right hand cube we would find ourselves inside the left hand one.

This is the basis of an amusing story by Robert Heinlein entitled *And He Built a Crooked House.* A man in Southern California (where else!) has a house

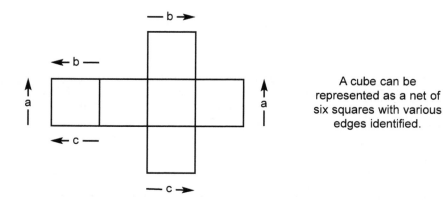

A cube can be represented as a net of six squares with various edges identified.

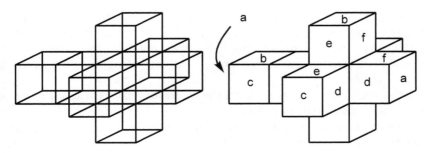

The net of a hypercube.

built for him one day by a strange builder. The house has eight nice rooms oriented as shown below. The five rooms on the main floor are a central hall, kitchen, dining room, lounge, and drawing room. There is a garage basement beneath the hall and master bedroom above the hall and finally a superb study in the attic above the master bedroom with wonderful views on all four sides. But when he and his wife come to move in they find only a single room, the basement. However, the builder takes them into the basement and surprisingly the other rooms seem to be all there but connected by strange staircases. And when they attempt to go out onto the roof of the study in the attic they find to their horror that they are back in the basement!

Well of course the house has folded up into a hypercube as the result of a minor earthquake. As a consequence, the rooms lead into each other in a most disconcerting manner. For example, if we attempt to go out of the master

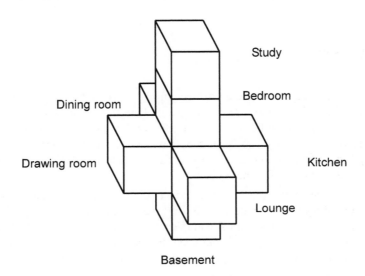

The Crooked House described by Robert Heinlein.

bedroom window above the kitchen we actually find ourselves falling into the kitchen from the ceiling. If we then attempt to leave the kitchen via the window opposite the central hall we find ourselves entering the study by the corresponding window and upside down! The story unfolds into a series of strange disasters.

If this seems all rather peculiar then consider the analogy with the cubic net of six squares which might be shown to a Flatlander.

A rather more sombre example of the hypercube net of eight cubes appears in the painting by Dali entitled *Crucifixion ('Hypercubic Body')*. This shows a man crucified on a cross taking the form of the hypercube net.

Further reading

FOR CONSIDERATION of the fourth dimension as being of mystic significance see for example *Theosophy and the Fourth Dimension* by Alexander Horne. For an advanced account of the geometry of regular figures in four dimensions see *Regular Polytopes* by Coxeter. For an introduction see *Introduction to Geometry* also by Coxeter. The honeycombs are discussed in *Mathematical Recreations and Essays* by Rouse Ball and *Mathematical Snapshots* by Steinhaus.

The story *And He Built a Crooked House* by Robert Heinlein will be found in *Fantasia Mathematica* compiled and edited by Clifton Fadiman. The painting by Dali is illustrated in *Dali* by de Liaño.

See also Appendix C for more ramblings about the fourth dimension and Appendix D for stereo images of projections from four into three dimensions.

Exercises

1. How many different colours are required to colour the six honeycombs in three dimensions? Each solid has every face the same colour. Solids with adjacent faces must be of different colours. Solids that meet only along an edge or at a vertex can have the same colour. The honeycomb of cubes requires only two colours, alternate cubes could be black and white. Now consider the other five honeycombs.

2. In two dimensions the length of the two diagonals between the opposite corners of a square of side 1 is $\sqrt{2}$. In three dimensions the length of the four diagonals between the opposite corners of a cube of edge 1 is $\sqrt{3}$. What is the length of the superdiagonals between the opposite corners of a hypercube of edge 1? How many such superdiagonals does the hypercube have?

3. In the crooked house, where do we find ourselves if we go through the four walls (imagine a door in each), the ceiling and floor (imagine hatches in

them) of the kitchen? Fill in the chart below. Give which surface of the new room you enter by and your orientation assuming you were standing vertical as you left the kitchen. The first row has been completed and shows that if you were to go down through a hatch in the floor, then you would be in the basement and have entered it through a wall and since you went through the floor feet first then you come through the wall feet first as well and so are lying down.

4* The Schläfli symbols in three dimensions using 3 and 4 are {3, 3}, {3, 4}, {4, 3} and {4, 4}. What are they? Observe that {4, 4} is actually a honeycomb (the square tiling). It is a fact that a sequence representing a honeycomb cannot appear as a subsequence in a higher dimension. Thus a sequence such as {3, 4, 4} is impossible. List all the combinations of 3 and 4 that make up valid Schläfli symbols in four and five dimensions and say what they are. Deduce that there can only be one honeycomb in five or more dimensions.

Face	Location	Enter by	Orientation
Floor	Basement	Wall	Lying down
Ceiling			
Wall opposite hall			
Wall adjacent to hall			
Wall to right of hall			
Wall to left of hall			

4 Projective Geometry

THE GEOMETRY we did at school (or maybe didn't) was mostly very dull. It was all about lengths and angles; proofs were often about showing that certain angles or lengths were equal. However, there is much geometry in which the lengths of lines and sizes of angles are not considered at all. This so-called projective geometry was heavily studied in the nineteenth century but became unfashionable. This was perhaps because it seemed to have no practical value and did not provide a foundation for other things. Nevertheless, it has a certain elegance and beauty.

Pappus' theorem

PAPPUS LIVED IN ALEXANDRIA in the fourth century and wrote an extensive treatise on mathematics in about 320. His famous theorem is a good starting point for a tour of projective geometry. Here it is.

Consider two lines l and l' and take any three points on each. We can call them A, B, C and A', B', C'. Now draw a line from each point on one line to the two points on the other line with a different letter. So through A we draw two lines, one to B' and one to C' and so on. The two lines joining A to B' and A' to B meet in a point which we can call C''. In a similar way the two lines BC' and $B'C$ meet in A'' and finally AC' and $A'C$ meet in B''.

The remarkable thing is that A'', B'' and C'' always lie on another straight line l''. This is Pappus' theorem. It makes no mention of lengths or angles and so is in strong contrast to the theorems of Euclid.

Note that there are nine lines and nine points with three points on each line and three lines through each point. It is symmetries like this that give projective geometry its great beauty.

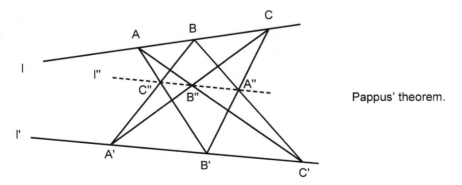

Pappus' theorem.

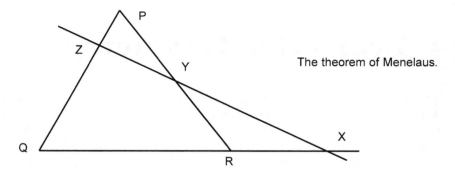

The theorem of Menelaus.

In a sense this was the beginning of projective geometry although Pappus did not know this and proved his theorem using traditional methods. The theorem of Menelaus who also lived in Alexandria but somewhat earlier (around AD 100) is about the ratios of lengths cut on the sides of a triangle by a fourth line. It is thus a typical traditional theorem. It is shown above. The product of the three ratios cut by the fourth line on the three sides is 1 thus

$$\frac{QX}{RX} \times \frac{RY}{PY} \times \frac{PZ}{QZ} = 1$$

In the example, Y is the midpoint of PR so that $PY = RY$ and Z is a quarter of the way from P to Q so that $QZ = 3 \times PZ$. It then follows from the formula that QX is 3 times RX.

It is possible to prove Pappus' theorem by using Menelaus' theorem five times on one of the triangles in the Pappus figure (e.g. the triangle formed from the lines $A'C$, $C'B$ and $B'A$). We leave this nasty exercise to the dedicated reader.

Pappus' theorem is stated in terms of points and lines and the only relationships we need are the Propositions of Incidence in a plane which are

> two points define a line,
>
> two lines define a point.

Note the symmetry in these two statements; this symmetry underlies the principle of duality which we met in the previous two lectures. If we exchange point for line and vice versa throughout then the propositions are unchanged.

There are similar propositions in three dimensions which are

> two points define a line,
>
> three points (not on a line) define a plane,
>
> two lines in a plane define a point,
>
> two lines though a point define a plane,
>
> three planes (not through a line) define a point,
>
> two planes define a line.

In three dimensions the symmetry is obtained by interchanging plane and point whereas line remains unchanged in the middle. The principle of duality in three dimensions thus means that any theorem involving points, lines and planes implies a corresponding theorem involving planes, lines and points.

It is not possible to prove Pappus' theorem using just these propositions and this gives the theorem deep significance. By contrast, the famous theorem of Desargues can be proved just from the propositions of incidence.

Desargues' theorem

GIRARD DESARGUES (1593–1661) was an architect and military engineer from Lyons. He is generally considered to be the founder of projective geometry and is famous for the following theorem concerning triangles.

Suppose two triangles ABC and $A'B'C'$ are such that the lines AA', BB' and CC' are concurrent which means that they go through a common point P. (Concurrent is from the Latin *con*, together, plus *curro*, I run.) It is then said that the two triangles are in perspective from P. Suppose now that sides BC and $B'C'$ meet in L and that CA and $C'A'$ meet in M and that AB and $A'B'$ meet in N. Then the theorem says that the three points L, M and N are collinear which means that they lie on a common line. This is sometimes expressed by saying that if two triangles are in point perspective then they are also in line perspective.

If the triangles are not in the same plane then the theorem is almost obvious since the three points L, M and N all lie on the line of intersection of the two planes containing the triangles. However, we do first have to show that the pairs of lines such as BC and $B'C'$ actually intersect. But this follows from the fact that both lines BC and $B'C'$ lie in the plane defined by the two lines PCC' and PBB'.

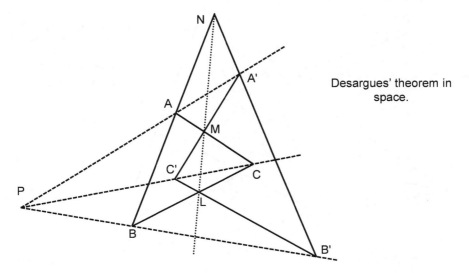

Desargues' theorem in space.

If both triangles are in the same plane then the proof is not so easy. In fact we have to introduce a third triangle outside the plane which is in perspective with both original triangles and we then apply the theorem of three dimensions which we have just proved.

Thus in the diagram below, the triangles ABC and $A'B'C'$ in black lie in the plane of the paper and then the line PQQ' in red is any line not in the plane of the paper and Q and Q' are any two points on it. The point A'' is where the red lines $A'Q'$ and AQ meet and so on for B'' and C''. We then have a triangle $A''B''C''$ which is shown in blue. It is then the case that the pair of triangles $A'B'C'$ and $A''B''C''$ are in perspective from Q' whereas the pair of triangles ABC and $A''B''C''$ are in perspective from Q.

Since ABC and $A''B''C''$ are in perspective it follows that BC and $B''C''$ meet on the line where the plane of the triangle $A''B''C''$ meets that of ABC. Applying the same argument with triangles $A'B'C'$ and $A''B''C''$ it follows that in fact BC, $B'C'$ and $B''C''$ all meet on this line and so this is the point L. Ditto M and N. So L, M and N all lie on the line where the plane of the triangle $A''B''C''$ meets the original plane. This proves it. The argument is perhaps not easy to follow but see Appendix B which has stereo images of these figures which might help.

It is strange that Desargues' theorem is so easy to prove in three dimensions but cannot be proved in two dimensions without the help of the third dimension. And indeed it is possible to construct special geometries in two dimensions in which the propositions of incidence hold but Desargues' theorem is not true. We will have a brief look at some special geometries in a moment.

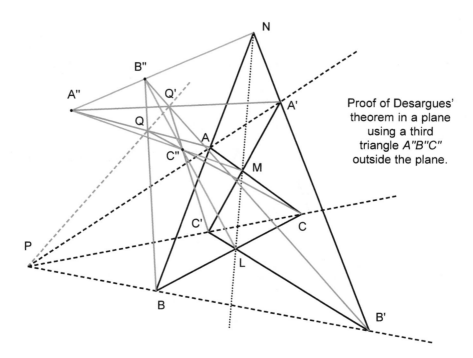

Proof of Desargues' theorem in a plane using a third triangle $A''B''C''$ outside the plane.

Duality

IT IS INTERESTING to consider the duals of Pappus' and Desargues' theorems in both two and three dimensions.

Before doing so it is worth noting how important it is to get a neat and symmetric notation. We are very used to talking about triangles by naming their three vertices. In fact the very word triangle emphasizes that they have three angles rather than three sides. So we talk about a triangle with vertices A, B and C. But we can equally talk about a triangle in terms of its three sides and it is convenient to denote them by lower case letters so that the sides are the lines a, b and c where the line a is the line opposite the point A and so on. We can then refer to a line either by its own symbol (such as a) or in terms of two points on it (such as BC). Equally, we can refer to a point either by its own symbol (such as A) or in terms of two lines through it (such as bc). It should now be clear that the dual of a triangle considered as three points is simply a triangle considered as three lines. So the whole concept of a triangle is self-dual. I suppose we should really refer to it as a trilateral when we think of it as three lines.

We can now phrase Desargues' theorem in the form: If two triangles ABC and $A'B'C'$ are such that the lines AA', BB' and CC' all go through a point X then the points aa', bb' and cc' all lie on a line x. The point aa' is of course the point L in the previous diagrams.

To obtain the dual of Desargues' theorem in two dimensions we have to interchange point and line throughout. We then get: If two triangles abc and $a'b'c'$ are such that the points aa', bb' and cc' all lie on a line x then the lines AA', BB' and CC' all go through a point X.

The interesting result is that we obtain the same theorem in reverse. So the converse is the dual theorem. This is sometimes expressed as saying that Desargues' theorem in two dimensions is self-dual.

We can do the same with Pappus' theorem. The original theorem is (see page 89) as follows. Consider two lines l and l' and take any three points A, B, C on l and A', B', C' on l'. Denote the point of intersection of BC' and $B'C$ by A'' and define B'' and C'' similarly. Then A'', B'' and C'' all lie on a line l''.

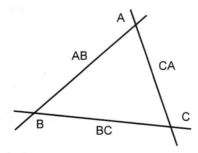

A triangle as three points ABC
joined by lines BC, CA and AB.

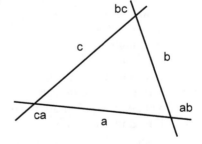

A triangle as three lines abc
meeting in points bc, ca and ab.

94 Gems of Geometry

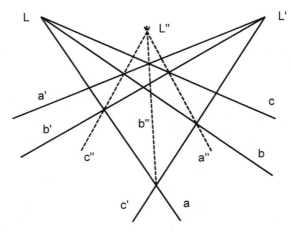

The dual of Pappus' theorem.

In order to obtain the dual we simply interchange point and line and so on. The result is as follows.

Consider two points L and L' and take any three lines a, b, c through L and a', b', c' through L'. Denote the line joining bc' and $b'c$ by a'' and define b'' and c'' similarly. Then a'', b'' and c'' all go through a common point L''. This is shown in the diagram above.

At first sight the dual of Pappus' theorem doesn't look like the original at all. But closer inspection shows that it really is the same diagram.

The original diagram has nine points A, B, C, A', B', C', A'', B'', C'' and also nine lines l, l', l'', BC', $B'C$, CA', $C'A$, AB', $A'B$. Moreover, each line has three of the points on it and each point has three of the lines through it.

The dual diagram also has nine points and nine lines and three of each through or on each of the other as appropriate. In order to show that the diagrams really are the same we need to show how the points of one can be identified with those of the other. This can be done in many ways but one way is to identify L with A and L' with A'. The three lines through L then correspond to l, AC' and AB' whereas the three lines through L' correspond to l', $A'B$ and $A'C$. Following the intersections of these lines we find that the lines a'', b'' and c'' correspond to l'', $B'C$ and BC'. These three lines intersect in A'' which is therefore the point corresponding to L''.

So in conclusion, the dual of Desargues' theorem in two dimensions is the same theorem in reverse and the dual of Pappus' theorem also gives rise to the same configuration although the identification is less clear.

The Pappus configuration has nine lines and nine points with three points on each line and three lines through each point. By contrast the Desargues configuration has ten lines and ten points with again three points on each line and three lines through each point. The charm of both configurations partly lies in their dual nature.

Duality in three dimensions

WE WILL NOW take a deep breath and consider the dual of Desargues' theorem in three dimensions. We recall that in this form of duality we have to interchange point and plane but leave line unchanged.

Desargues' theorem can be stated as: If two triangles ABC and $A'B'C'$ are such that the lines AA', BB' and CC' all go through a point X then the points aa', bb' and cc' all lie on a line x. We assume that the triangles are in different planes.

Before considering the dual it will be helpful to understand the amazing symmetry of the configuration. In fact there are five planes: the two planes of the triangles ABC and $A'B'C'$ and the three planes involving the perspective point X which are $XBB'CC'$, $XCC'AA'$ and $XAA'BB'$.

The Desargues configuration is indeed simply that caused by the intersection of five planes. Two planes meet in a line and since there are ten ways to choose two objects from five (thus 12, 13, 14, 15, 23, 24, 25, 34, 35, 45) we get ten lines. Three planes meet in a point and since there are also ten ways to choose three objects from five, we get ten points. Each of the planes contains four of the lines (one corresponding to each of the four other planes) and each such line contains three of the points.

In order to appreciate the symmetry the figure is shown below with the lines labelled l_{12} and the points labelled P_{12} and so on. Each line contains the three points whose suffices are different from those of the line; thus line l_{14} contains the points P_{23}, P_{25} and P_{35}. Similarly, point P_{14} is on the lines l_{23}, l_{25} and l_{35}.

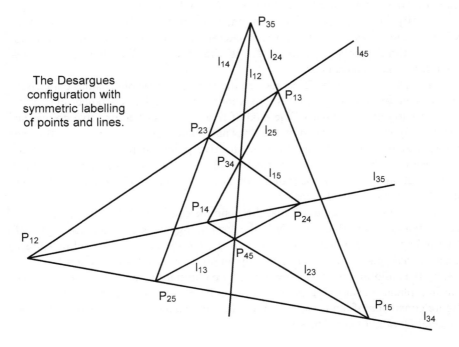

The Desargues configuration with symmetric labelling of points and lines.

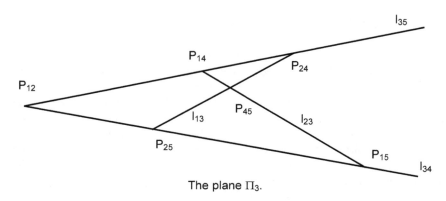

The plane Π_3.

We can call the five planes Π_1, Π_2, Π_3, Π_4, Π_5 where we number them so that the plane Π_1 contains the six points without a 1 in a suffix plus the four lines with a 1 in the suffix. The line l_{12} is then the line of intersection of the planes Π_1 and Π_2, and the point P_{12} is the point of intersection of the three planes Π_3, Π_4, and Π_5.

Because the diagram is so symmetric it should now be clear that it reveals Desargues' theorem in lots of different ways. We can take *any* of the ten points as the point of perspective of two triangles and then the line with the same suffices is the line of perspective of the theorem. Thus if we take P_{24} as the point of perspective then the triangles concerned are $P_{12}P_{32}P_{52}$ and $P_{14}P_{34}P_{45}$ and their pairs of sides meet in the three points P_{13}, P_{15} and P_{35} which lie on the line l_{24}.

Finally, note that each plane contains four lines and six points. These form what is known as a complete quadrilateral in which each of the four lines contains three points. The plane Π_3 is shown above. We shall discuss the complete quadrilateral later.

In conclusion then the Desargues configuration in three dimensions comprises that formed by five planes and is thus known as a complete pentahedron.

Now to return to duality. The dual of five planes is of course five points. So the dual figure is that formed by the lines and planes through five points. But the thought of five points brings to mind the five points of the 4-simplex in four dimensions which we met in Lecture 3. So the dual of the Desargues configuration can equally be thought of as the projection of a 4-simplex in four dimensions onto three dimensions.

Now we saw in Lecture 3 that the 4-simplex (or 5-cell) has five points, ten edges, ten faces and five cells. Or in the terminology of this lecture, the figure of five points in four dimensions has ten lines, ten planes and five solids (three-dimensional spaces). If we slice this by an arbitrary solid then the ten lines are cut by the solid in a point, the ten planes are cut in a line and the five solids are cut in a plane – everything goes down one dimension. So the cross-section is a figure consisting of ten points, ten lines and five planes – in other words it is the original Desargues configuration!

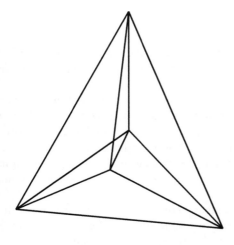

The simplex of five points has ten lines and ten planes and five solids.

So at the end of the day the Desargues configuration and its dual are simply cross-sections and projections respectively of the simplex in four dimensions consisting of five arbitrary points and the lines and planes joining them.

The dual of Pappus' theorem in three dimensions is less interesting. This is largely because (unlike Desargues' theorem) it is really only a theorem of two dimensions so duality in other dimensions is rather artificial.

We have seen that we have duality in a plane by interchanging point and line and duality in three dimensions by interchanging point and plane. Other dualities are possible.

For example, if a configuration just consists of lines and planes through a common point then a different form of duality can be obtained by interchanging line and plane. This results in another configuration consisting of planes and lines through a common point. The reason this works is that if we take the three-dimensional propositions of incidence and consider what they reduce to if all the lines and planes go through a point then we simply get

> two lines (though a point) define a plane,
>
> two planes (through a point) define a line.

and these reveal the dual nature of planes and lines though a point. Note that they correspond to the usual duality of line and points in a plane and so the whole thing is consistent.

Similar dualities can be created in four dimensions. The normal duality in four dimensions transforms point into solid and line into plane and vice versa. Dualities can also be formed concerning all the solids, planes and lines through a point and so on.

I think that's quite enough about duality and trust that the reader will have gained some idea of the power of the technique. I hope it has illustrated some of the beauty of this kind of geometry but maybe it has been just all too confusing.

Infinity and parallels

IT IS PERHAPS time to say something about the fundamental nature of projective geometry. Basically, it concerns those geometrical properties which are unchanged by a projection.

For example, if we draw the Pappus diagram on a window and then observe the shadow cast upon the floor by the sun, the shadow will equally be a Pappus diagram. However, if we draw the diagram of Pythagoras' theorem which comprises a right-angled triangle with squares on its sides on the window, then the shadow on the floor will be distorted; the triangle will no longer be right-angled and the squares will no longer be squares. So Pythagoras' theorem is not a theorem of projective geometry.

An important consideration in Euclidean geometry is the business of parallel lines. As we know, parallel lines do not meet which singles them out from other pairs of lines. But parallelism is not a projective property. A painting of parallel railway tracks in the usual perspective manner shows them as pairs of lines which meet. Indeed, as we remarked at the beginning, projective geometry is not about angles and lengths because these get changed by projection.

In projective geometry we do not contemplate the existence of parallel lines at all and simply say that all pairs of lines meet in a point. This makes life much easier because it introduces the duality between points and lines because all pairs of points define a line and so it is reasonable that all pairs of lines should define a point.

The reader might feel that projective geometry must be useless if it does not admit of real-world matters such as parallel lines. However, we can change any projective theorem into a Euclidean one by introducing the so-called line at infinity. Since we want parallel lines to meet somewhere, we introduce the idea of points at infinity where they meet. There is one point at infinity in each direction and we say that these points together lie on a line which is "at infinity". Note carefully that a pair of parallel lines only meet at one point at infinity. Both directions lead to the same point at infinity. I suppose we can think of the line at infinity as a sort of infinite circle where the two ends of any diameter are identified as the same point – we will encounter this idea of identification again in the lecture on Topology.

Now we can take any projective theorem and decree that any one line in it is the line at infinity. It then becomes a corresponding Euclidean theorem in which the line at infinity does not appear and any lines meeting on the line at infinity are now considered to be parallel.

The rails meet
at infinity.

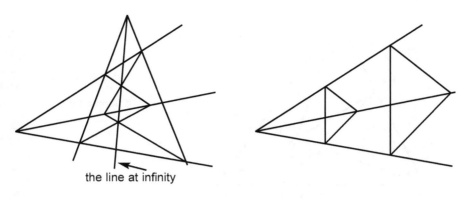

the line at infinity

Let us try this with Desargues' theorem. Suppose that the line at infinity is the line where the pairs of sides meet. The diagram on the left above shows the original Desargues configuration and that on the right shows the configuration with the line at infinity removed and those pairs of lines meeting at infinity now drawn as parallel. The new diagram shows that if two triangles are such that their sides are parallel then the triangles are in perspective from some point.

Quadrilaterals and Quadrangles

WE MENTIONED the complete quadrilateral when discussing the duality of Desargues' theorem. A complete quadrilateral is the figure composed of four lines; these four lines meet each other in a total of six points. The dual figure is a complete quadrangle which is the figure composed of four points; these four points join each other in six lines.

The six points of the quadrilateral are arranged as opposite pairs which we can denote by P_{12}, P_{34}, by P_{13}, P_{24}, and by P_{14}, P_{23} where we use the convention that P_{ij} is the point of intersection of lines l_i and l_j. (Note the importance of using a consistent notation. It would be very hard to follow this if we called the lines a, b, c, d and the points A, B, C, D, E, F.)

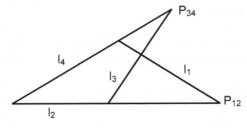

A complete quadrilateral has four lines.
A complete quadrangle has four points.

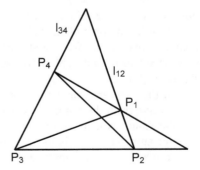

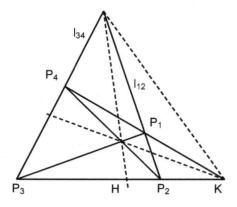

A quadrangle showing its diagonal triangle.

In a dual manner the six lines of the quadrangle are also arranged as opposite pairs which we can denote by l_{12}, l_{34} and so on.

Now we can introduce what is known as the diagonal triangle of these figures. We will concentrate on the quadrangle and leave it to the reader to construct the dual argument for the quadrilateral if desired.

Each opposite pair of lines of the quadrangle meet in a point which is known as a diagonal point. Since there are three opposite pairs of lines there are therefore three diagonal points and these form a triangle known as the diagonal triangle of the quadrangle. In the diagram above the diagonal triangle is shown dashed.

An important property of the diagonal triangle is that its sides cut the lines of the quadrangle in pairs of points such as those marked H and K on the line joining P_2 and P_3. Note that K is of course one of the points of the diagonal triangle. The four points P_2, P_3, H and K are said to form a harmonic range. In fact if we fix P_2, P_3 and K and then draw any quadrangle to fit around them it turns out that H is always at the same point (this can be proved using Desargues' theorem).

Further insight can be gained by supposing that the diagonal line through K and the point of intersection of l_{12} and l_{34} is the line at infinity. It then follows that the original quadrangle is a parallelogram because its opposite sides meet at infinity. Thus the side P_2P_3 is parallel to P_1P_4 and P_1P_2 is parallel to P_3P_4. The point H is then the midpoint of the side P_2P_3.

In order to come clean over this topic it is necessary to briefly introduce the idea of cross-ratio. We noted earlier that lengths are not fixed in projective geometry and in fact neither are ratios. But rather strangely the ratios of certain ratios are fixed.

In the diagram opposite we say that the ranges $ABCD$ and $A'B'C'D'$ are in perspective from the point P. Clearly the lengths of the segments AB and $A'B'$ are not the same. Nor are ratios such as AB/BC the same as $A'B'/B'C'$. However, perhaps surprisingly, it is the case that the ratio of the ratios AB/BC and AD/DC *does* remain the same whatever the projection.

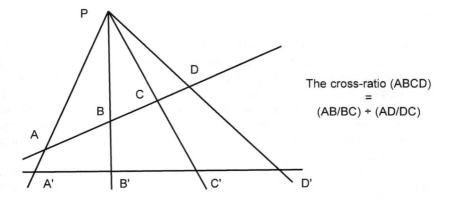

The cross-ratio (ABCD)
=
(AB/BC) ÷ (AD/DC)

This can be proved using old-fashioned geometry and in particular the sine rule. Remember that in any triangle *ABC*, the sines of its angles are in the same proportion as the lengths of the opposite sides.

We can then apply this rule to various triangles with apex at *P*. After some messing about we find that

$$\frac{AB}{BC} \bigg| \frac{AD}{DC} = \frac{\sin APB}{\sin BPC} \bigg| \frac{\sin APD}{\sin DPC}$$

This may look a bit indigestible but all that has happened on the right is that each segment on the left has been replaced by the sine of the angle that segment subtends at the point *P*. The remarkable thing about the expression on the right is that it is the same whether we use the points *ABCD* or the points *A'B'C'D'* since the angles are the same in both cases. That proves that the so-called cross-ratio of four points remains the same under projection. So despite the fact that lines and angles change size under projection here we have found a numerical property that does not change. Incidentally, a group of lines such as *PA*, *PB*, *PC*, *PD* through a point *P* is known as a pencil of lines through *P*.

Now back to the harmonic range that we found on the quadrangle. A harmonic range is a range in which the cross-ratio is exactly −1. (It is negative because we measure some of the distances backwards.) So in other words the points *H* and *K* are such that they separate P_2 and P_3 in the same ratio internally

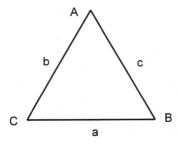

The sine rule:

$$\frac{\sin A}{a} = \frac{\sin B}{b} = \frac{\sin C}{c}$$

and externally. And so when we send K off to infinity the ratios both become 1 and so H is the midpoint of P_2P_3.

The four points of a cross-ratio can be considered in six different ways. That's because having fixed one of the four points the other three can be arranged in 3! = 6 ways thus $ABCD$, $ABDC$, $ACBD$, $ACDB$, $ADBC$, $ADCB$. If the value of one cross-ratio is x, then the others are $1/x$, $1-x$, $1/(1-x)$, $(x-1)/x$ and $x/(x-1)$. These six values can be transformed into one another by the two operations of subtracting from 1 and taking the inverse (dividing into 1). In the case of the harmonic range the six values become -1, -1, 2, 1/2, 2, 1/2 and we see that they reduce to three values only in this special case.

Well that is quite enough about cross-ratio for the moment. We will meet it again when we consider bubbles in Lecture 6 but I wanted to make the point that projective geometry does have some numerical properties.

Conics

SO FAR we have only considered points and lines but geometry concerns much more. Indeed, Euclid has a lot to say about circles and his later books consider conic sections generally.

In classical times the conic sections were defined in terms of the sections of a cone. Thus if we cut an ordinary cone parallel to its base we get a circle. If we cut it at a small angle we get an ellipse. If we increase the angle (and assuming that we have a proper double cone), then the cross section breaks into the two parts of a hyperbola. Just before this happens the section is a parabola. Finally, if we cut the cone through its vertex then we get a pair of straight lines which shows that a pair of lines can also be treated as a special kind of conic section.

There are various other ways of defining conics other than as sections of a cone. For example we can define an ellipse as those points for whom the sum of the distance from two other points is a constant. The two other points are the foci of the ellipse. The orbit of the planets about the Sun is an ellipse with the Sun at one focus. If the foci coincide then the ellipse becomes a circle.

In projective geometry we do not distinguish between the different kinds of conics because they can be projected into one another. If we draw a circle on a piece of glass and project the image onto the floor then we will typically get an ellipse but by varying the position of the point of projection (or that of the floor!) we can get the other forms of conic as well. In fact this is easily seen by considering the double cone once more; we simply take the vertex of the cone as the point of projection and the circle as a section of the cone. So in projective geometry all conics are the same and by convention we usually draw them as an ellipse. But of course the pair of straight lines as a form of conic remains distinct.

An interesting theorem concerning conics is that named after Blaise Pascal (1623–1662), the famous French mathematician who sadly died at a relatively early age. Pascal's theorem is a generalization of Pappus' theorem and is most

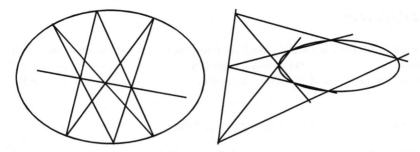

Two illustrations of Pascal's theorem.

neatly stated as follows. If a hexagon be inscribed in a conic then the points of intersection of the opposite sides of the hexagon are collinear. If we make the hexagon zigzag about then we get a diagram very like that of Pappus where the line of the three concurrent points crosses the conic. Alternatively we can have other diagrams where the line lies outside the conic as shown above.

We can obtain a Euclidean theorem by taking the line of collinearity as the line at infinity. We then deduce the theorem that if a hexagon is inscribed in a conic and two pairs of opposite sides are parallel then the third pair are also parallel.

Duality with respect to conics is interesting. We normally think of a conic as a set of points. The dual concept is, of course, to think of a conic as a set of lines and in fact the lines concerned are simply the tangents to the conic. Using this idea we can now find the dual of Pascal's theorem. The dual of a line joining two points on a conic is the point of intersection of two tangents to the conic. We quickly see that the dual theorem is as follows. If the sides of a hexagon are the tangents of a conic then the lines joining opposite vertices are concurrent. This was discovered by C J Brianchon (1760–1854) when he was a student at the Ecole Polytechnique in Paris and so is known as Brianchon's theorem. These were early days of understanding duality and one could get a theorem named after one for doing it.

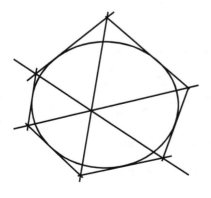

Brianchon's theorem. If the sides of a hexagon are the tangents of a conic then the lines joining opposite vertices are concurrent.

Coordinates

MUCH ELEMENTARY GEOMETRY is about coordinates and equations concerning them. A general point (in a plane) typically has coordinates (x, y) giving the distance of the point from two fixed axes and the equation of a line is usually stated as

$$y = mx + c \qquad \text{equation of a straight line}$$

We haven't used coordinates so far and at first sight they might appear out of place because they would seem to imply the measurement of lengths and we know that lengths have no place in projective geometry.

However, a very beautiful system of homogeneous coordinates can be devised where we use three coordinates (x, y, z) to represent a point (we are discussing the geometry of a plane) rather than the usual two coordinates. So a point might have a position such as $(1, 2, 1)$. Moreover, the system is such that it is only the ratios of the values that matter so that the point $(1, 2, 1)$ can equally be called the point $(2, 4, 2)$. An important rule is that there is no point $(0, 0, 0)$.

The line joining two points P and Q is then defined as consisting of all those points whose coordinates have the form $\alpha P + \beta Q$ where α and β are any numbers

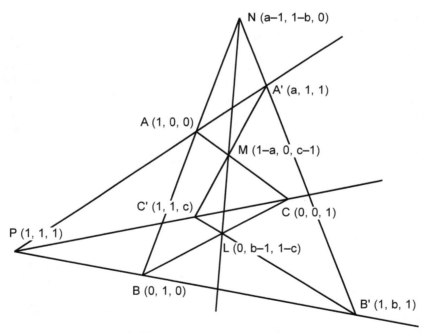

Proof of Desargues' theorem using coordinates.

The point of perspective P is the unit point.

at all. For example, if P is the point (1, 0, 1) and Q is the point (0, 1, 2) then taking $\alpha = 1$ and $\beta = 2$ we obtain the point $P+2Q$ which is (1+2×0, 0+2×1, 1+2×2) which is (1, 2, 5). So the point (1, 2, 5) lies on the line joining P and Q.

Another basic principle is that we can choose any three points (not on the same line) as the points with coordinates (1, 0, 0), (0, 1, 0) and (0, 0, 1). These form the so-called triangle of reference. We can then choose any other point (not on the triangle) as the unit point (1, 1, 1).

This system of so-called homogeneous coordinates is very powerful and has a certain symmetric beauty. As a simple example we can consider Desargues' theorem once more. Suppose that the triangle ABC is the triangle of reference so that A is the point (1, 0, 0), B is (0, 1, 0) and C is (0, 0, 1). We can take the point of perspective P to be the unit point (1, 1, 1). See the diagram opposite.

Now any point on the line PA will have coordinates which are a combination of those of P and A (remember the general form $\alpha P + \beta Q$ introduced above). And so A' must have the form $(\alpha+\beta, \beta, \beta)$. But it is only the ratio that matters so we can equally and quite generally consider the point A' to be $(a, 1, 1)$. In the same way we can take B' as $(1, b, 1)$ and C' as $(1, 1, c)$.

Now the point L is where BC meets $B'C'$ and so must be a combination of B (0, 1, 0) and C (0, 0, 1) and also a combination of B' (1, b, 1) and C' (1, 1, c). Since it has to be a combination of B and C, it follows that its x-coordinate is 0. It should then be clear that the combination of B' and C' required has to be $B'-C'$ so that the x terms cancel. It follows that L is the point (0, $b-1$, $1-c$).

In the same way M is the point $(1-a, 0, c-1)$ and N is $(a-1, 1-b, 0)$. Now we find that $L+M+N = 0$ or in other words $L = -M-N$ and this shows that L is on the line joining M and N and so in other words L, M and N lie on a line. So we have proved Desargues' theorem using coordinates.

Using coordinates in projective geometry has a certain beauty because of its intrinsic symmetry. With a suitable choice of triangle of reference we can make things easy because of this symmetry.

As we have seen, a theorem in projective geometry can be turned into a Euclidean one by choosing one line to be at infinity. It is a common convention to choose the line $z = 0$ as the line at infinity; that is the line whose points all have the third coordinate equal to zero. For the other points, if we then scale the coordinates so that the third coordinate is 1, then the x- and y-coordinates become the normal Euclidean ones.

Finite geometries

AND NOW to a rather curious topic. It was mentioned earlier that there are special geometries with strange properties. We are so used to our experience of the real world that we take it for granted that all lines have an infinite number of points on them. However, we can construct geometries in which this is not the case and in which every line has only a finite number of points and dually there

106 Gems of Geometry

are only a finite number of lines through each point. These finite geometries still satisfy the various rules such as the propositions of incidence and so the appropriate theorems still hold true.

Finite geometries were extensively studied by Gino Fano (1871–1952), an Italian mathematician; in 1892 he discovered the smallest finite geometry in which there are only seven points and lines and where each line has only three points and vice versa. It is usually represented by the diagram below. Unfortunately it is not possible to represent all the lines in this geometry with what appear to be straight lines. Six lines seem straight but the seventh looks rather like a circle! Moreover, the only points are those shown as big blobs. The circular line might look as if it meets all the other lines but it doesn't. There are just the seven points and the seven lines shown.

We can give coordinates to the Fano plane as shown in the diagram. However, the coordinates can only be 0 or 1. Moreover, arithmetic is done modulo 2 which means that we cast out multiples of 2 and only consider the remainder. So $1 + 1 = 0$ in this arithmetic.

We now find that if we add the coordinates of any two points then we do indeed get the third point of the line through them. Thus if we add (0, 1, 0) to (1, 0, 0) we get (1, 1, 0) which is indeed the other point on that line. Similarly, if we add the point (1, 1, 0) to one of the others such as (0, 1, 0) then we get (1, 2, 0) which of course reduces to (1, 0, 0) since $2 = 0$ in this arithmetic and again this is the other point on that line. So the relationship of adding the points is quite symmetric. By adding the coordinates we are simply using the general formula $\alpha P + \beta Q$ that we introduced earlier. However, in this geometry, because of the modulo 2 arithmetic, the only possible values of α and β are 0 and 1.

We cannot illustrate Desargues' theorem using the Fano plane because we cannot find two triangles in perspective. Thus taking (0, 1, 0) as the point of perspective, and (1, 0, 0), (1, 1, 1), (0, 0, 1) as one triangle, then the remaining three points do not form a triangle because they are the three points of the "circular" line.

Another example of a finite geometry is the 13-point geometry of the plane which has four points on each line.

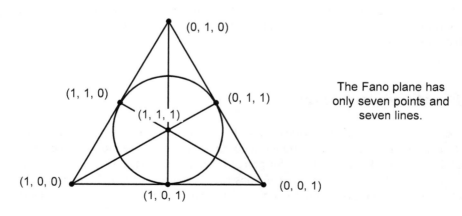

The Fano plane has only seven points and seven lines.

The coordinate system of this geometry is based on modulo 3 arithmetic which is such that 1+2 = 0 and 2+2 = 1.

The additional six points have a two in their coordinates and are 102, 021, 210, 112, 121, 211. (For compactness we will omit the parentheses and commas in the coordinates in future – no confusion should arise because the individual coordinates are single digits.) Other combinations such as 122 are covered by the six mentioned. Thus 122 is equivalent to 244 which reduces to 211 since 4 = 1 in this arithmetic. Similarly 201 is the same as 102. Point 102 is on the line joining 100 to 001 and point 112 is on the line joining 001 and 111 and so on.

It becomes difficult to represent this 13-point geometry with a diagram. The Fano geometry was awkward since one line became a "circle". In the 13-point geometry things are much worse. One representation is shown below with some lines in colour and dotted and dashed which hopefully helps to distinguish them.

Now we can try out Desargues' theorem. Take for example the two triangles (110, 121, 021) and (100, 111, 001). They are easily seen to be in perspective from the point 010. Now consider where the corresponding sides meet. The side (110, 121) meets the side (100, 111) at the point 011. Similarly the other two sides meet at the points 112 and 101. And then we find that 011, 112 and 101 lie on the same line. So Desargues' theorem does work. We leave it to the reader to try out Pappus' theorem in the same way.

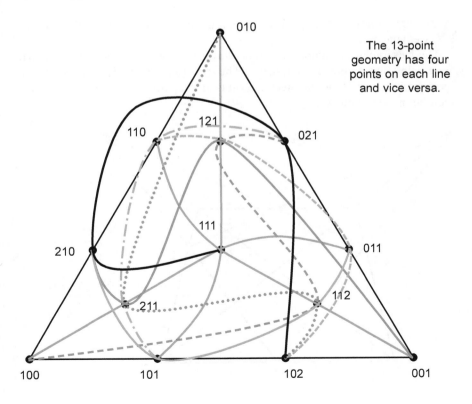

The 13-point geometry has four points on each line and vice versa.

Many finite geometries of this kind are possible. In fact, for every number n which is a power of a prime number, there is a geometry which has n^2+n+1 points and lines with $n+1$ points on each line and vice versa. (Note that n has to be a power of a prime number including a prime number itself. It does not work for 6, 10, 14 and so on because these are products of different primes. The reason concerns group theory which is outside the scope of these lectures.)

The Fano plane corresponds to $n = 2$ and has $4+2+1 = 7$ points and lines, and $2+1 = 3$ points on each line. The case $n = 3$ is the 13-point geometry which we have just discussed with $9+3+1 = 13$ lines and points, and 4 points on each line.

Finite geometries extend to three or more dimensions as well. The number of points in three dimensions is n^3+n^2+n+1. By duality this is also the number of planes whereas the number of lines is $(n^2+1)(n^2+n+1)$. So the Fano space where $n = 2$ has 15 points and planes and 35 lines with 7 points in each plane.

Incidentally, the things we call points and lines can be considered as quite abstract. All that matters is that there is a relationship between them corresponding to "is on" and that two of one uniquely define one of the other in some way. See the story in Appendix G.

Configurations

WE WILL CONCLUDE with a few points regarding some of the configurations we have met.

The Desargues configuration can be drawn in many different ways and that shown below has an elegant symmetry. The triangles ABC and $A'B'C'$ are again in perspective from P and the meeting points of the pairs of sides are L, M and N which lie on a straight line.

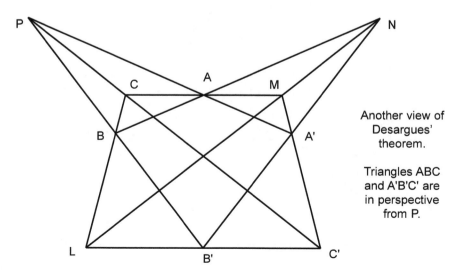

Another view of Desargues' theorem.

Triangles ABC and A'B'C' are in perspective from P.

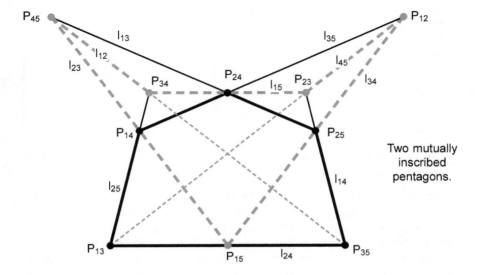

Two mutually inscribed pentagons.

The diagram can be relabelled using P_{ij} and l_{ij} as before. Two other curious properties are worth noting.

In the figure above, the lines $l_{12}, l_{23}, l_{34}, l_{45}, l_{15}$ are red and dashed. Thus five lines are red and five lines are black. They form two pentagons and the vertices of the red pentagon lie on the black lines (extended as necessary) and vice versa.

In the figure below, the lines $l_{15}, l_{25}, l_{35}, l_{45}$ are red and dashed. The four red lines form a complete quadrilateral and the six black lines form a complete quadrangle. Each point of the quadrilateral (there are six, namely, $P_{12}, P_{13}, P_{14}, P_{23}, P_{24}, P_{34}$) lies on a line of the quadrangle (P_{12} on l_{34} and so on).

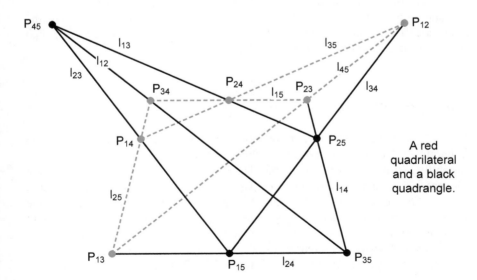

A red quadrilateral and a black quadrangle.

Configuration	Points	L/point	Lines	P/line
Triangle	3	2	3	2
Quadrangle	4	3	6	2
Quadrilateral	6	2	4	3
Fano plane	7	3	7	3
Pappus	9	3	9	3
Desargues	10	3	10	3
13-point geometry	13	4	13	4

Coxeter mentions these two elegant properties of the Desargues configuration in Exercises 2 and 3 in Section 3.2 of his *Projective Geometry*.

The Pappus configuration has a number of amazing properties as well. For example, having taken three points on one line, the three points on the other line can be ordered in six different ways and so six different Pappus lines can be drawn. These form two groups of three concurrent lines. In the case of the six points on the conic in Pascal's theorem, there are sixty different ways of considering the points to form the successive vertices of a hexagon and so there are sixty Pascal lines. The resulting figure is known as the Hexagrammum Mysticum of Pascal. It occupied the minds of many Victorian geometers.

Configurations in a plane can be classified by giving the number of lines l and points p in them. Note that if the number of points on each line is p_l and the number of lines through each point is l_p then we always have

$$l \times p_l = p \times l_p$$

since they both represent the number of pairings in the configuration.

The configurations we have met are tabulated above. Note that the quadrilateral and quadrangle are duals whereas all the others are self-dual including the familiar triangle.

Further reading

IT IS HARD to know quite what to recommend as a next step since there is a vast literature. The little book by Faulkner entitled *Projective Geometry* is a possibility. Of a more advanced nature is *Projective Geometry* by Coxeter or the relevant parts of his *Introduction to Geometry*. For the Hexagrammum Mysticum consult *Plane Geometry* by H F Baker – this is volume 2 of his comprehensive treatise on geometry which is very hard and hard to find. A gentler treatment including the proof of Pappus' theorem using Menelaus' theorem will be found in *Geometry Revisited* by Coxeter and Greitzer.

See also Appendix B for stereo images of the Desargues configuration.

Exercises

1. Two points define a line and two lines define a point. In the Fano plane, the two points 100 and 010 define the line to the left of the diagram. The two points 011 and 101 define the line that looks like a circle. At which point do these two lines intersect?

2. Let A, B and C be any points on a line. Suppose A', B' and C' are points on another line such that AB' and $A'B$ are parallel and that CB' and $C'B$ are parallel. Using Pappus' theorem and choosing a suitable line at infinity deduce that AC' and $A'C$ are also parallel. See the diagram below.

3*. Prove Pappus' theorem using coordinates in a similar way to the proof of Desargues' theorem. Take the points as follows $A = (a, 1, 0)$, $B = (b, 1, 0)$, $C = (c, 1, 0)$, $A' = (a', 0, 1)$, $B' = (b', 0, 1)$, $C' = (c', 0, 1)$. (So that ABC is the line $z = 0$ and $A'B'C'$ is the line $y = 0$.) Then show that the point whose coordinates are $(aa'-bb', a'-b', a-b)$ lies on the line AB'. Deduce that it also lies on the line $A'B$ and so must be the point C''. Similarly, find the coordinates of A'' and B'' and show that the coordinates are such that a combination of those of A'' and B'' are the coordinates of C'' and thus they are on a straight line.

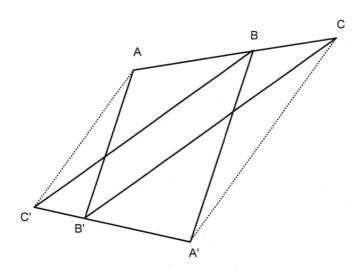

5 Topology

TOPOLOGY is one of those branches of mathematics where some of the simple results are very easy to state and appreciate. However, the underlying mathematics is rather hard and it is not easy to prove what seems quite obvious. Topology used to be considered a branch of geometry although now it is seen more as a branch of abstract algebra. In this lecture we will just look at some of the entertaining facts and not generally bother with how they might be proved.

Hairy dogs

THIS TOPIC traditionally has this title since it can be postulated as: "Can you comb your dog flat without any awkward whorls?" Well a dog is basically a sphere with bits sticking out such as legs so we can consider an abstract formulation of the problem.

Imagine a ball covered in hair. Can we comb it flat? The answer is No. Try as we might there are typically two places where the hair will not lie flat. We might get a sort of vortex or a point where all the hair leads away or all comes together.

On the other hand consider a rubber ring (like an inner tube). Now imagine such a ring covered with hair. Can we comb this flat? Yes we can. We can comb the hair round and round the short way or we can comb it the long way round and either way it lies flat.

So there is a real difference between a sphere and a torus (to give the ring its proper name) in this respect. The study of this sort of property is known as

A hairy sphere and a hairy torus.

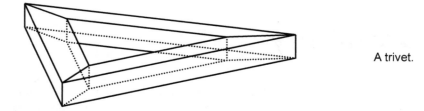

A trivet.

topology from the Greek τοπος (*topos*), a place. Topology concerns the way in which surfaces and shapes are connected together but it does not concern lengths and angles and similar properties. Thus in topological terms a sphere is the same as a cube or tetrahedron and a torus is the same as a teacup.

We may recall that the regular figures in three dimensions satisfy Euler's formula

$F + V - E = 2$ Euler's characteristic for sphere

where F is the number of faces, V is the number of vertices and E is the number of edges. In topology, where the rigid shape does not matter, the regular figures can be thought of as representing maps on the surface of a sphere and an important topological property of a sphere is that the Euler characteristic is two.

However, consider the hollow figure shown above (such as a wooden trivet for standing hot saucepans upon). This has 24 edges, 12 vertices and 12 faces so that $F+V-E$ is zero. The hollow figure is topologically a torus and in fact the Euler characteristic for a torus is zero. So here we have a real numerical difference between the sphere and the torus.

It is the Euler characteristic which controls whether the hairy dog can be combed smooth. We can comb the hairy torus smooth because its Euler characteristic is zero. But in the case of the sphere there are typically at least two places where the hair is troublesome. The troublesome places (called singular points) can be characterized in various ways. They might be places where the hair all flows away or where it all comes together; other possibilities are a vortex in one direction or the other; all of these score +1. Another form of singular point

Sources, sinks and vortices score +1.

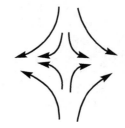

Simple saddles score −1.

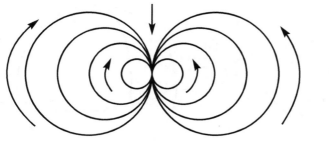

A dipole scores 2.

is where the hair flows in from North and South and flows out from East and West or vice versa with a stagnant point in between much like the contours of a mountain pass or saddle; such a point scores −1. A further form of singular point is a dipole which in essence is the merger of a source and a sink and so scores 2; we will look at this in more detail in a moment.

So the rule is that the sum of the scores of the singular points must equal the Euler characteristic and so be 2 on a sphere and zero on a torus.

As an obvious example, we can comb a sphere so that the hair flows away from the North pole (a source) and towards the South pole (a sink) and this confirms the score of 2. We could equally comb it following the lines of latitude and this gives a vortex at both poles and therefore a score of 2 once more.

A more curious example is provided by a dipole. Consider a line that is a tangent at the North pole of a sphere and then consider the series of planes containing that line. These planes will intersect the sphere in circles and it is clear that we can comb the sphere smoothly along these circles. At first sight it might appear that this enables us to comb the complete sphere smoothly. However, the situation at the pole itself is curious. As the planes become more nearly tangential to the sphere, the circles of intersection become smaller and smaller as in the pattern above. If we approach the pole from a side then the hair is all neatly lying one way until suddenly at the pole itself it is the other way. It's a very curious mess.

The directions of flow around the dipole are similar to the lines of force around a short bar magnet. They flow out of the North pole and into the South pole (these are magnetic poles not poles on a sphere). One magnetic pole is a source and the other is a sink. If the source and sink move together then ultimately we get a dipole and so score 2.

As another example consider a hairy octahedron as shown overleaf. If we comb the hair in a circular form but in opposite directions in adjacent faces then we find that there is a vortex in the centre of each face and there is a saddle around each vertex. Since there are 8 faces and 6 vertices, the total score is clearly $F-V = 8-6$ which is 2.

It should be observed that in any map where two lines cross at each vertex (that is four edges meet at each vertex), then the number of edges E must be twice the number of vertices V. So the formula $F+V-E$ in this case becomes $F-V$

Hairy octahedron.

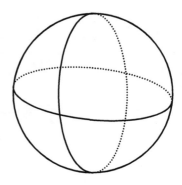

Equivalent map on sphere.

as well. Note that in order to be able to comb a figure in the way we did the octahedron, each vertex must have an even number of faces meeting at it. We can call such vertices crosspoints because two or more lines cross at them. If two lines cross then we get a simple form of crosspoint with score -1. If three lines cross then it scores -2 and so on.

The octahedron is clearly equivalent to a sphere with equator and two meridians. Although the octahedron is the simplest figure with four flat faces meeting at each vertex, a simpler map on the sphere is just with equator and one meridian in which case there are four edges, four faces and two vertices which again satisfies Euler's formula.

Note that a sphere with just one great circle such as the equator is not a proper map since the one edge does not join any vertices – but just adding one vertex in the edge somewhere makes it satisfactory. Another restriction for a map to be allowed is that a face must not have a hole in it. Thus if we removed the lines on the top and bottom flat surfaces of the trivet then two faces would have holes in them and the Euler formula would not apply.

The torus can be thought of as a sphere with a handle on it. The most general surface of this kind is a sphere with p handles on it. The Euler characteristic in this general case is $2-2p$. An example of the case of two handles is obviously a two-handled vase; another example is a teapot. A pretzel has three handles and so has Euler characteristic -4. So a hairy pretzel certainly cannot be combed smooth!

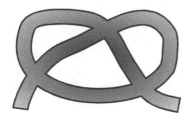

A pretzel is a sphere with three handles.

Colour problems

ANY MAP on a plane or a sphere can be coloured with just four colours without sections sharing an edge having the same colour. This has long been conjectured but was only proved as recently as 1976 by Kenneth Appel and Wolfgang Haken. An interesting feature is that the proof was partly done by a computer program and so an air of strangeness hangs over it. Mathematical proof is an odd business. To a large extent it is a discourse between the writer and the readers. The writer needs to convince the readers that the proof is correct. Typical mathematical proofs proceed by stages in which some "obvious" intermediate steps are omitted. This does not matter since the readers can fill them in as necessary. However, in the case of a proof by computer, the readers really have to convince themselves that the computer program is perfect and this is not easily done. Nevertheless, it seems that the four colour theorem is true.

Strangely enough the corresponding theorem on the torus that seven colours are required was easily proved long ago. We can illustrate the four colour theorem by considering the colouring of some of the figures we met in earlier lectures.

The tetrahedron requires the full four colours since each face meets every other face and it has four of them. However, the coloured tetrahedron exhibits enantiomorphism since there are two ways of doing the colouring. The two versions are mirror images. Thus in the diagram below there are two tetrahedra with red, blue, green and yellow faces and they are both standing on their yellow face.

This property of the tetrahedron is important in organic chemistry since the carbon atom is quadrivalent. The carbon atom can be thought of as being at the centre of the tetrahedron with its four bonds pointing to the vertices. So if a compound has four different atoms or groups attached to the central carbon atom

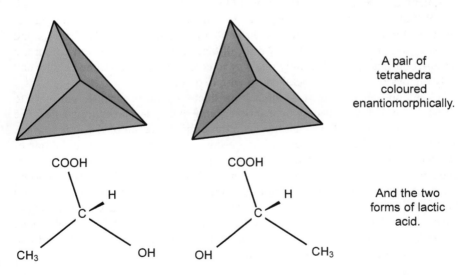

A pair of tetrahedra coloured enantiomorphically.

And the two forms of lactic acid.

Gems of Geometry

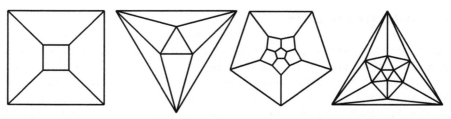

Schlegel diagrams for the cube, octahedron, dodecahedron and icosahedron.

then the compound can exist in two different forms. This phenomenon was first observed by Louis Pasteur (1822–1895) the great French chemist. He showed that lactic and tartaric acid exist in two forms. It is curious that living material often favours one form whereas synthetic processes usually produce both forms. A tragedy occurred when the drug thalidomide which reduced morning sickness in pregnant women in one form caused deformities in the foetus in the other form.

Returning to geometry, the cube can be coloured with three colours since the opposite faces can be coloured the same. The octahedron only needs two colours – it is unique amongst the five regular solids in having an even number of faces meeting at each vertex and so these can be coloured alternately. The icosahedron needs three colours but the dodecahedron needs all four colours. The cuboctahedron is another example of a figure only needing two colours.

An interesting way to represent solid figures is to imagine them opened out so that one face extends to infinity in all directions and the others are then projected upon it. This is essentially the view we would obtain by looking at a transparent model from a close viewpoint. Such representations of the cube, octahedron, dodecahedron and icosahedron are shown above. The viewpoint chosen is on the circumsphere. They are called Schlegel diagrams after the German mathematician Victor Schlegel (1843–1905).

We can use the Schlegel diagrams to illustrate the colouring of the regular figures. One face is represented by the rest of the plane outside the diagram and we shall assume it to be yellow. We recall that there are two ways to colour the tetrahedron. There is in fact only one way to colour a cube with three colours and only one way to colour an octahedron with two colours.

The dodecahedron is more interesting since there are four different ways to colour it and these form two enantiomorphic pairs as shown opposite. The dodecahedron has 12 faces and there are three faces of each of the four colours. Moreover, each face of a given colour (yellow, say) is opposite a different one of the three other colours. (The 12 faces form 6 pairs and there are 6 ways of choosing two colours from four.) Thus in the diagrams we have chosen to orient the dodecahedra so that the background yellow face is the one opposite the red face which thus appears in the middle. The other two yellow faces which clearly have to be adjacent to the central red face are opposite a green and blue face respectively. The other faces around the red face are either two green and one blue or vice versa and this accounts for the four arrangements.

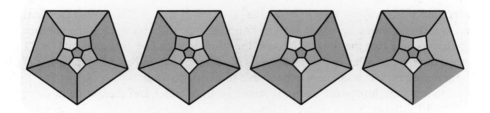

Four ways of colouring a dodecahedron.

The case of the icosahedron is rather complex and there are in fact no fewer than 144 different ways in which it can be coloured using the three colours that are necessary. It would take too long to explain these and so the interested reader should consult the references.

We will now briefly consider the tilings of the plane which, as we saw in Lecture 2, are like infinite solid regular figures. The familiar chessboard illustrates that the square tiling requires only two colours. The triangular tiling also needs only two colours but the hexagonal tiling requires three. Of the mixed tilings, only {3.12.12} requires the full four colours; the reader may recall that this tiling is obtained by taking the hexagonal tiling and filing down the corners so that triangles appear and the hexagons then become dodecagons. Since the same basic hexagonal colouring is needed for the dodecagons it follows that a fourth colour is required for the triangles.

A related puzzle is how many different ways are there to colour the regular figures using a different colour for each face? We already know that the answer for the tetrahedron is two. In the case of the cube the answer is thirty. To see this choose one of the six colours – we will suppose it is red. Paint one face red and place it with the red face down on the table. Then there are five different choices for the top face. We can then paint any side with one of the four remaining colours and turn it so that it faces North. The three remaining faces can then be

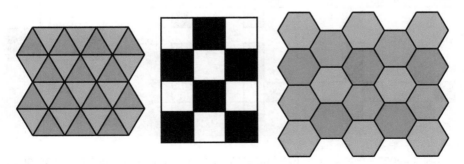

Tilings of triangles and squares require two colours, but hexagons require three.

120 Gems of Geometry

painted in the three remaining colours in six ways. (Recall that n items can be arranged in $n!$ ways where $n! = n \times (n-1) \times (n-2) \times \ldots \times 2 \times 1$.) So the final answer is five times six which is thirty. Another way to arrive at the same answer is first to note that the four colours on the four sides can be arranged in $4! = 24$ ways but then to observe that this results in some repetition because the cube can be oriented in four ways so we then have to divide by four.

It is a straightforward matter to make 30 little cubes and then to paint them in the 30 different ways. If we add one more so that two are the same then we can pose our friends the interesting challenge of finding the identical pair. This is straightforward if approached logically but many people just thrash hopelessly picking up arbitrary pairs and putting them back in disgust.

An interesting puzzle regarding these 30 cubes was devised by Major P A MacMahon (1854–1929). He was a major in the army and served in India and later became a famous mathematician. The puzzle is to take any one of the 30 cubes and then find 8 from the remainder that can be arranged to form a matching cube $2 \times 2 \times 2$ where not only do the external faces match the original but the touching faces of the individual cubes match as well. The 8 cubes can be chosen in only one way but curiously they can be arranged in two ways. The diagram below shows one arrangement – the four on the left form the bottom layer, those in the midde form the top layer and the large cube itself is on the right. The nets represent the cubes in an obvious way; thus the large cube has a white base, green top, yellow back, red front, black left, and blue right face. The reader is invited to look for the other arrangement.

There is also the question of how many ways are there to colour the octahedron, dodecahedron and icosahedron so that all faces are different. We can use the same approach as with the cube. For a general regular figure with n faces, having chosen a colour for the bottom face we can choose the top face in $n-1$ ways. The remaining $n-2$ colours can then be applied to the sides in $(n-2)!$ ways.

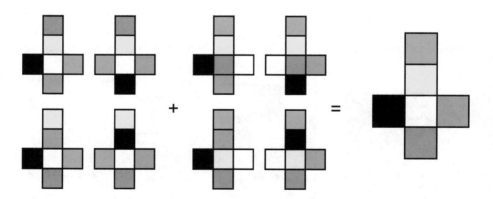

Eight cubes which can be fitted together to form a large cube with six different external faces and matching touching internal faces.

We then have to divide by the number of edges around the base (k, say) in order to account for duplication caused by the rotational symmetry. So we end up with

Number of ways to colour n-hedron with k-gon faces = $(n-1)! / k$.

This formula also works for the tetrahedron. We get

Tetrahedron	$3!/3 = 2$
Cube	$5!/4 = 30$
Octahedron	$7!/3 = 1680$
Dodecahedron	$11!/5 = 7983360$
Icosahedron	$19!/3 = 40548366802944000$

It is clearly just feasible but very tedious to make a set of 1680 coloured octahedra but the others are obviously out of the question.

Colouring maps on the torus

IN ORDER to explore properties of figures such as the torus it is convenient to introduce the idea of an identification model rather like the nets we discussed when talking about the crooked house in the lecture on four dimensions.

If we take a rectangle and imagine the two long sides joined together, we get a cylinder. If we then join the two ends of the cylinder together in the obvious way, we finally get a torus. We can indicate these connections by rectangles with arrows showing the edges joined together as shown below.

We can now use the model of the torus to show that a map requiring seven colours is possible. The map is simply a hexagon surrounded by six other hexagons and because of the way in which the torus is connected each of the

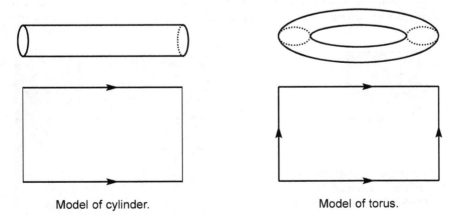

Model of cylinder. Model of torus.

122   Gems of Geometry

A torus with a map of seven hexagons; each one is surrounded by all the others.

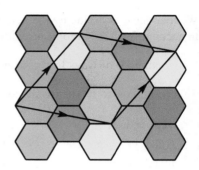

A section of the hexagonal tiling of the plane showing the derivation of the torus.

hexagons is surrounded by all the others so they all touch each other and therefore must have different colours. This map is really just a section from a hexagonal tiling of the plane as shown above; of course the tiling in the plane doesn't need seven colours but it is coloured the same way for comparison. Many other hexagonal maps on the torus are possible but this is the only one that needs the maximum of seven colours. Note that the model of the torus has been skewed so that the hexagons look regular. Incidentally, these maps are called regular if every face has the same number of edges and so on – the faces don't have to be regular figures with equal sides and angles and in any event they are usually not flat.

Square maps on the torus are also possible. The map shown below is intriguing in having five squares with each square being surrounded by the other four. Such square maps again relate to the square tiling of the plane. Another example of a square map on the torus is provided by the trivet which we met earlier; this map has 12 squares. The reader might like to consider how many colours are required to colour the map on the trivet.

Note that the usual representation of the torus is effectively a map of one square with just one vertex. That may sound very peculiar; it also has only two

A torus with a map of five squares; each one is surrounded by all the others.

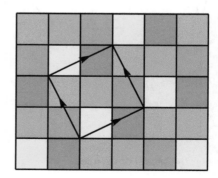

edges because the opposite edges are the same. Nevertheless it satisfies the Euler formula that $F+V-E$ is zero on the torus.

Percy J Heawood (1861–1955), a mathematician from Durham, conjectured that the maximum number of colours required for a map on a surface which is a sphere with p handles is given by the following peculiar formula

$$N = \text{integer part of } \frac{7 + \sqrt{48p+1}}{2} \quad \text{Heawood's formula}$$

For the sphere ($p=0$), this formula gives the correct value of 4, for the torus ($p=1$), it gives 7. In the case of a teapot ($p=2$), the expression $(7 + \sqrt{48p+1})/2$ becomes 8.424... and so we take the whole number part which is 8. In the case of the pretzel ($p=3$), the answer is 9 and so on.

Although Heawood's formula has proved to be correct, he was unable to prove it in the basic case of mapping a sphere or plane. As mentioned earlier that had to wait until the advent of computers enabled a tedious proof to be mechanized.

It is perhaps a more amazing fact that on these surfaces, for all values of p, there is a regular map of N regions each of which has $N-1$ sides and meets all the others. We have seen that in the case of the sphere ($p=0$), the tetrahedron provides a map of four triangles. On the torus ($p=1$), we have seen the map of seven hexagons. So a teapot ($p=2$) requires 8 colours for a map of eight heptagons all meeting each other and a pretzel ($p=3$) requires 9 colours for a map of nine octagons.

The Möbius band

IF WE TAKE a strip of paper and glue the ends together in the usual way then we get a cylinder. If, however, we put a half turn in the strip and then glue the ends together we get a strange surface known as the Möbius band (or strip) which is named after the German astronomer August Möbius (1790–1868).

The Möbius band can be represented by the diagram below right in which the fact that the two edges are joined after a twist is indicated by the arrows going in opposite directions.

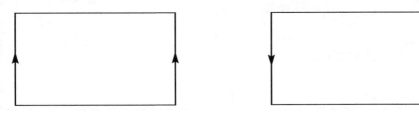

Model of cylinder.   Model of Möbius band.

124   Gems of Geometry

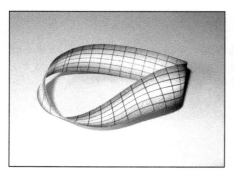 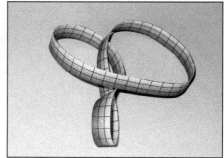

Cutting a Möbius band in half.

The key property of the Möbius band is that it has only one side and only one edge. This can easily be seen by simply running a finger along it. The Möbius band figures in two woodcuts by Escher entitled *Möbius Strip I* and *Möbius Strip II*. The former shows a band with a cut down the centre. The latter shows red ants crawling over the surface of a Möbius band.

If we cut a Möbius band down the centre (see above) and then open it out we find that we still have a single strip but it is twice as long and has two full turns in it. Because they are full turns it has two edges and two sides and is topologically equivalent to a cylinder; one side is the original side of the Möbius band (grey) and the other side is the cut we made (white). The fact that it has turns does not matter topologically and if we lived in four dimensions we could in fact undo the turns without cutting it. If we now cut the strip with two turns down the middle then, being equivalent to a cylinder, it naturally becomes two pieces. However, they both have double turns and are in fact linked together.

Another experiment is to cut the Möbius band down its length but one-third of the way across as shown below. Because of the one-sided nature of the band the cut goes around twice and finally joins up. This results in two linked pieces:

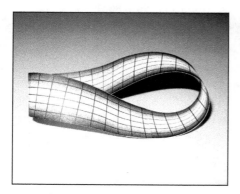 

Cutting a Möbius band in thirds.

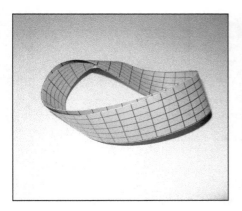 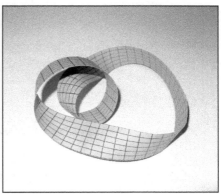

A double Möbius band joined up and then shaken out.

one is a Möbius band of the same length as the original and the other is a doubly twisted cylinder twice as long. The new Möbius band comes from the centre of the old one (and is white) and its one edge is the new cut, the twisted cylinder comes from the outside of the original band with one edge being that of the original band and the other being the new cut.

This perhaps unexpected behaviour of the Möbius band is the basis for a short story by William Upson entitled *Paul Bunyon versus the Conveyor Belt* in which Loud Mouth Johnson loses two bets regarding what happens when a conveyer belt is slit down the middle in order to make it longer.

A third trick is to take two equal strips of paper flat against each other, make a half twist in them and then join the two pairs of ends together. (In the picture above the two strips are red one side and green on the other.) This seems like two Möbius bands snuggled together and indeed one can run a finger all around between them. However, on shaking it out, it is revealed as just a single piece and in fact is another cylinder with a double twist. It is an interesting exercise to turn the twisted cylinder back into the snuggled bands.

It is not so surprising that we get a cylinder with a double twist since the double Möbius band is just what we would get by slicing a thick Möbius band in two. And that is obviously much the same as slicing a strip of rectangular cross-section with a half twist and clearly it doesn't matter in topological terms whether the rectangular section is split along the long or short side; splitting along the long side is like cutting the Möbius band down the middle.

The Möbius band and the cylinder both have Euler characteristic of zero. This is easily seen by considering the representation of them as a single rectangle with two ends identified. In both cases this results in a map which has one face, two vertices and three edges so that $F+V-E$ is zero. Since the Euler characteristic is zero it is possible to comb a hairy cylinder and a hairy Möbius band quite smooth and indeed it is pretty obvious that we can do this.

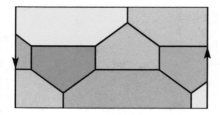 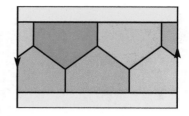

Two maps on the Möbius band needing six colours.

Maps on the Möbius band may require six colours. Two examples are shown above: the one on the left has three pentagons and three hexagons ($F=6$, $E=18$, $V=12$) whereas the one on the right has five pentagons and one hexagon ($F=6$, $E=16$, $V=10$). The hexagons are the faces with an edge on the boundary; each face is adjacent to all five others and so six colours are required. These two maps are closely related as we shall see later. Note that when colouring a one-sided surface we have to think of the colour penetrating right through so that both "sides" at every point have the same colour.

Both the Möbius band and the cylinder may be found in a torus. That this is so is best seen by considering a torus held horizontally and then inserting a knife vertically so that it intersects the axial circle. (The axial circle is the circle which goes all the way around in the middle.) We then move the knife all the way around keeping it vertical and following the axial circle. The result is that the torus is divided into two parts and the new face created by the knife is a cylinder. On the other hand consider what happens if, as we move the knife around the torus, we steadily change the angle of the knife so that when we are one half of the way, it is horizontal and when we get back to the beginning it is vertical again but reversed so that the cut joins up. It will seem as if we must have cut the torus into two pieces but this is not so. The cut is a Möbius band and the torus remains intact. This is exactly like the experiment of gluing the two twisted strips together that we mentioned earlier – they seem as if they must be two Möbius bands but in fact they are just one cylinder with a double twist.

The Möbius band and the cylinder are examples of open surfaces which means that they have free boundaries. The torus and sphere are closed surfaces.

The Möbius band and cylinder have some things in common such as that the Euler characteristic is zero for both. However, they are quite different regarding maps. Maps on the Möbius band can require six colours as we saw above whereas maps on the cylinder are similar to maps on the sphere and only require four colours. In order to prove this, consider the section of a sphere between the arctic and antarctic circles; this is obviously a cylinder. We can then shrink the polar circles until they vanish in which case the cylinder becomes a sphere. It is clear that any map on a cylinder requiring more than four colours would then become a map on the sphere requiring more than four colours but we know that this is not possible. So all maps on the cylinder require at most four colours.

The Klein bottle

THE KLEIN BOTTLE is a very strange surface. It is named after its discoverer, the German mathematician Felix Klein (1849–1925). Perhaps the best way to introduce the Klein bottle is through the identification model shown below. It is similar to that for the torus except that one pair of edges is matched in the opposite direction. It can be constructed by taking a cylinder and joining its two free ends in the opposite way to when making a torus but in order to do this the cylinder has to be deformed so that it intersects itself as shown. The Klein bottle might therefore seem somewhat improper because of the intersection but if we were lucky enough to live in four dimensions then we could make a Klein bottle that did not suffer from this apparent defect.

The Klein bottle is another example of a one-sided surface like the Möbius band and this accounts for a number of its peculiar properties. Despite being a closed surface, both inside and outside are the same. It features in a number of light-hearted tales such as *The Last Magician* by Bruce Elliott.

The Euler characteristic of the Klein bottle is zero like that of the torus and so a hairy Klein bottle can be combed smooth as is fairly obvious. Unlike the torus, only six colours are required for maps on the Klein bottle. As an example consider the first of our two six-coloured maps on the Möbius band. If we identify the two free edges it becomes a map of three pentagons and three heptagons on the Klein bottle. In the second diagram overleaf this has been emphasized by shifting the dividing line between the three heptagons so as to explicitly reveal the additional interfaces. This is a very peculiar map since each of the three heptagons meets both of the others twice as well as meeting the three

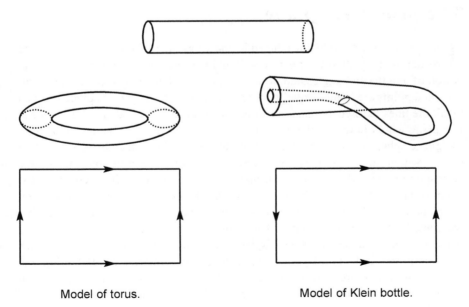

Model of torus.   Model of Klein bottle.

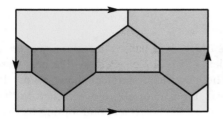 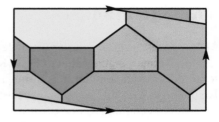

Two views of a map of three pentagons and three heptagons on a Klein bottle.

pentagons. As a consequence there are three points at which the three heptagons all meet.

Observe that the other map on the Möbius band does not become a proper map on the Klein bottle because the yellow face then goes all the way around the bottle and divides it into two parts. As a result the yellow face cannot be shrunk away because it has a hole in it (we say that it is not simply connected).

Before moving on, it should be noted from the rectangular identification model that rather than starting with a cylinder and identifying the two free edges in the opposite direction, we could alternatively start with a Möbius band and then identify the free edges in the same direction. A complication is that the two seemingly free edges of the Möbius band are in fact the same edge as we know so it's not that easy – we shall come back to this in a moment.

The projective plane

THE FINAL SURFACE to be introduced is known as the real projective plane. The reader will recall from Lecture 4 that in projective geometry there is no notion of lengths and angles and no concept of parallelism; as a consequence all lines and points are treated equally. A good model of the projective plane is given by all the lines in three dimensions through a point (this is the dual of all the lines in a plane which in turn is the dual of all the points in a plane, hence the name). Imagine a sphere with centre at the point, then every line meets the sphere in two points which are antipodes. So there is a pair of points on the sphere corresponding to each point of the projective plane. However, if we consider just a hemisphere then we only have one point on the hemisphere for each point on the plane except that there are two points on the equator. In order to remedy this we consider the opposite points of the equator as the same. So the projective plane can be modelled as a hemisphere with opposite points on its equator identified. We can simplify this for ease of presentation by flattening the hemisphere into a circle so the projective plane then becomes just a circle with opposite points of the circumference identified. Gosh, I hope that's clear.

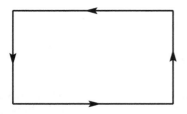
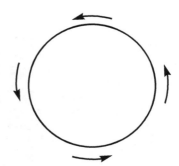

Models of projective plane.

Alternatively, we can model the projective plane using our familiar rectangular diagrams with some edges identified. In this case both pairs of edges have to be identified and both are reversed. This completes all the possibilities for identifying the pairs of opposite edges. Neither reversed gave the torus, one pair reversed gave the Klein bottle and now both pairs reversed gives the projective plane. Clearly, both the rectangular model and the circular model shown above are the same.

Six colours are required for mapping the projective plane and the example below shows six pentagons each of which is adjacent to all the others. This map is related to the two maps on the Möbius band which we saw earlier as will be explained in a moment. The map has 6 faces, 15 edges and 10 vertices and so the Euler characteristic, $F+V-E$, is 1. This is consistent with the obvious behaviour of a hairy projective plane as seen from the circular model; clearly we can smooth it around in a circular manner except for a single central vortex whose score is 1.

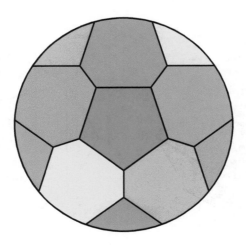

Map of six pentagons on a projective plane.

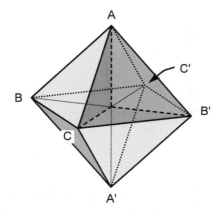

Three squares plus four triangles make a heptahedron.

In Lecture 2 we met a strange non-convex regular figure known as the heptahedron. This has three square faces and four triangular faces and it is easily seen that it is one-sided. In the diagram the opposite pairs of vertices are labelled A and A', B and B', and C and C'. So the three squares are $BCB'C'$ (blue), $CAC'A'$ (red), and $ABA'B'$ (green); the four triangles are ABC, $A'B'C$, $A'BC'$, and $AB'C'$.

If we unfold the heptahedron around the square $ACA'C'$ then we get the map shown below left where two of the squares have been distorted. Note how the opposite points have to be identified since they are the same point in the heptahedron. If we now slightly change the shape we arrive at the figure on the right in which it is perfectly clear that the opposite sides of the figure are identified in opposite directions. Thus we now see that the heptahedron is topologically equivalent to the projective plane. So the projective plane is another one-sided surface.

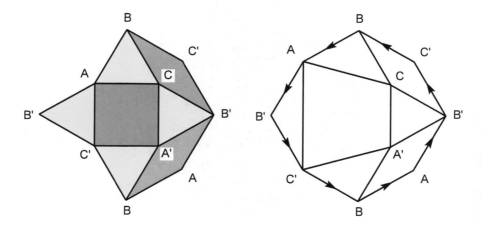

The heptahedron is an example of a projective plane.

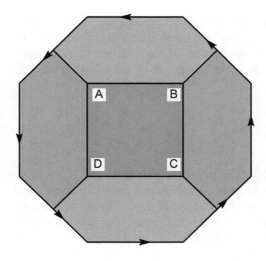

Three squares on a projective plane.

The heptahedron has 7 faces, 6 vertices and 12 edges thus confirming that its Euler characteristic is one. Since four faces meet at each vertex, it can be coloured with only two colours (like the cuboctahedron). The triangles can be one colour and the squares the other. The fact that the squares intersect each other does not matter.

Another interesting map on the projective plane consists of three squares each of which meets the others on two edges as shown above. There are only four vertices, A, B, C and D. The square $ABCD$ is red, the square $ACDB$ is green and the square $ADBC$ is blue.

In *Sylvie and Bruno Revisited* by Charles Dodgson (better known as Lewis Carroll), Mein Herr instructs Lady Muriel to join together three handkerchiefs to make the Purse of Fortunatus. First, she joins two together (the red and blue ones for example) and this results in a Möbius band. The one side of this band consists of four handkerchief edges and the next stage is simply to sew the green handkerchief edge by edge to these four edges. There are problems of course and

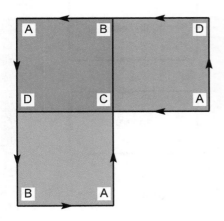

These are the three handkerchiefs and Lady Muriel has already sewn some edges together.

She then has no problem joining AD to DA so that the red and blue handkerchiefs form a Möbius band.

But completing the Purse of Fortunatus by joining the remaining three edges of the green handkerchief to the others is less easy.

she promises to do it later. Mein Herr says that it is called the Purse of Fortunatus because of its one-sided nature since everything outside the purse is also inside the purse and so the purse contains the wealth of the whole world. In four dimensions she could sew the handkerchiefs together without having to tear them.

Round up

WE HAVE NOW MET the main surfaces in two-dimensional topology. We have seen that some have two sides such as the sphere and torus whereas others are somewhat bizarre and only have one side such as the Klein bottle and projective plane. Most can be represented by identification diagrams as shown below. Note in particular the diagram for the sphere. Using single and double arrows clarifies which edges are identified. In most cases it is the opposite sides of the diagram but in the case of the sphere, the top edge is identified with the right side and the bottom edge with the left side. The poles can be thought of as at the top right and bottom left corners.

However, the identification diagrams do not cover the more curious surfaces such as the two-handled vase. In order to categorize all surfaces neatly it is convenient to introduce an awkward object known as a crosscap.

One way to introduce the crosscap is to take a Möbius band and consider how it would have to be rearranged so that its one edge could be neatly attached

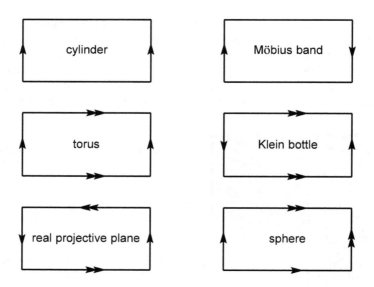

Summary of identification diagrams.

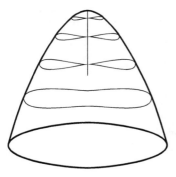

A crosscap.

to another free edge such as one end of a cylinder. The edge of the band is not knotted in any way so it ought to be possible to make it into a circle. However, we cannot do this without the band having to intersect itself. Well, we are familiar with the idea of a surface having to intersect itself from our experience of the Klein bottle and so it is not too awful an idea. The final result is the peculiar looking hat shape shown above.

The cross-section is a circle at the bottom and, as we progress upwards, the circle becomes pinched until it eventually becomes a figure-of-eight. The Möbius band intersects itself along the vertical line where the figures-of-eight cross over.

Now we have already observed that every two-sided closed surface is equivalent to a sphere with p handles. The torus had $p=1$ and so on. It can also be shown that every one-sided closed surface is equivalent to a sphere with one or more (q, say) discs cut out and replaced by crosscaps.

The case of one crosscap ($q=1$) gives the projective plane whereas two crosscaps ($q=2$) gives the Klein bottle. The Möbius band is equivalent to a projective plane with a disc cut out and this explains the relationship between the map of pentagons on the projective plane and the two maps of pentagons on the Möbius band. The first map on the Möbius band occurs if the hole in the projective plane is at an intersection between three faces whereas the map with yellow all along the free edge is obtained by cutting the hole in the projective plane in the middle of the yellow hexagon.

We have seen that the Euler characteristic X for a sphere with p handles is $2-2p$; in the case of a sphere with q crosscaps it is $2-q$, so we have

$X = 2 - 2p$ Euler characteristic for sphere with p handles,
$X = 2 - q$ Euler characteristic for sphere with q crosscaps.

The remarkable formula of Heawood for the colours on a map can be rewritten as

$$N = \text{integer part of } \frac{7 + \sqrt{(49 - 24X)}}{2} \qquad \text{Heawood's formula}$$

Object	Open/closed	Sides	Euler no.	Chromatic no.
Sphere	closed	two	2	4
Torus	closed	two	0	7
Teapot	closed	two	−2	8
Pretzel	closed	two	−4	9
Projective plane	closed	one	1	6
Klein bottle	closed	one	0	6
Cylinder	open	two	0	4
Möbius band	open	one	0	6

This formula also applies to one-sided surfaces except for one case and that is the Klein bottle. The formula predicts that seven colours are needed on the Klein bottle but in fact it has been shown that six colours always suffice. Note finally that Heawood's formula only applies to closed surfaces. It does not apply to open surfaces such as the cylinder and Möbius band.

The properties of the surfaces we have encountered are summarized in the table above. Note that N (the maximum number of colours required for a map) is sometimes called the chromatic number. Another point is that the proper technical terms rather than two-sided and one-sided are orientable and nonorientable.

Further reading

AN EASY INTRODUCTION to Topology will be found in *What is Mathematics?* by Courant and Robbins. There is a more difficult but very rewarding introduction in *Introduction to Geometry* by Coxeter. Somewhat easier perhaps is *Elementary Topology* by Blackett. The full details of colouring the icosahedron are given in *Mathematical Recreations and Essays* by Rouse Ball. The history and solution of the four colour problem are described in *Four Colours Suffice* by Robin Wilson.

The story *The Last Magician* by Bruce Elliott will be found in the anthology entitled *Fantasia Mathematica* compiled and edited by Clifton Fadiman. The story *Paul Bunyon versus the Conveyor Belt* by William Upton will be found in another anthology entitled *The Mathematical Magpie* also compiled by Clifton Fadiman. This latter anthology also includes the episode concerning the *Purse of Fortunatus* from *Sylvie and Bruno Revisited* by Charles Hodgson. These anthologies both contain other stories around the themes of topology.

Note also that Klein Bottles can be obtained from Acme Klein Bottles, 6270 Colby Street, Oakland, CA 94618, USA (www.kleinbottle.com).

See also Appendix D for more details of Schlegel diagrams.

Exercises

1. The hairy octahedron can be combed into swirls on each face with consistent hair direction along the edges because an even number of faces meet at each vertex. It is the only one of the five regular figures like this. Which of the 13 Archimedean figures have four faces at each vertex? (See Lecture 2.) Compute in each case the number of faces minus the number of vertices.

2. Draw the Schlegel diagram for the tetrahedron. Make a copy and colour the two diagrams differently (but with the same four colours of course).

3. How many colours are required to colour a trivet? Draw a representation of a trivet as a torus divided into twelve squares and then colour this.

4* On the model diagrams for a torus and Klein bottle (below) a representation of a cylinder is shown (the grey section) on the torus. Show that a cylinder and a Möbius band may be found in a Klein bottle by drawing similar representations. Finally, draw a Möbius band on the torus in a similar manner (use the knife cutting example as a guide).

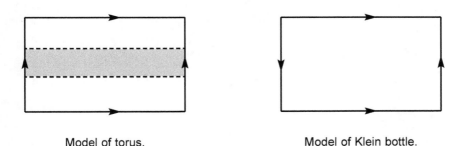

Model of torus.   Model of Klein bottle.

6 Bubbles

SOAP BUBBLES may seem an odd topic for a lecture. One might think that there was little to say since they are patently spherical and surely that is that. However, when two or more bubbles are joined together some intriguing features are revealed. This lecture is partly inspired by Boys' famous book entitled *Soap Bubbles* whose second edition was published in 1911.

Surface tension

IF WE BLOW UP a balloon made of rubber then the pressure increases as the balloon gets bigger. Eventually the molecules in the rubber are all straightened out and the stretching process turns into a big bang.

Soap bubbles are very different. As the bubble gets bigger, the pressure gets less. In order to show this let us recall the simple physics of surface tension.

Liquids behave as if they have a skin at the interface to the air. This skin has a tension which depends upon the liquid. We can imagine a line embedded in the skin and we define the surface tension in terms of the force acting at right angles on that line. If the force on a line of length one centimetre is one dyne then we say that the surface tension is one dyne/cm. (Remember that a dyne is the force required to accelerate a gram by one cm/sec/sec; it is very small – by contrast, the force of gravity upon a gram is 981 dynes.)

The surface tension of water is about 73 at room temperature, that of mercury is 465 and that of liquid gold is about 1100.

Although surface tension is very small for most purposes which immediately concern us, it is vital for many aquatic insects such as waterboatmen whose whole way of life depends upon being able to walk on water.

Now consider a soap bubble with surface tension T and consider the forces on half of it defined by a line which is a circumference around it. (See the diagram overleaf.) The film on one side of the line will pull on the line by a force which is $2T$ multiplied by the length of the line. (It is $2T$ because a film has two sides.) If the bubble has radius R then the length of the circumference is $2\pi R$ and so the force is $4T\pi R$. This is the force acting upon one half of the bubble and it is counterbalanced by the difference in pressure P between the air inside and outside the bubble. This difference acts over the area of the cross-section which is πR^2 and so the opposing force is $P\pi R^2$. Equating these two forces we get

$$4T\pi R = P\pi R^2 \quad \text{and so} \quad P = 4T/R$$

So the pressure in a soap bubble is inversely proportional to the radius of the bubble. Double the radius and the pressure is halved.

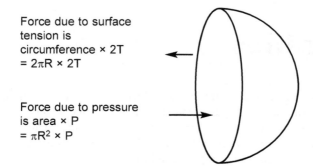

Force due to surface tension is
circumference × 2T
= 2πR × 2T

Force due to pressure is area × P
= πR² × P

The forces on half a bubble.

This gives some very unintuitive behaviour. Imagine two large glass tubes connected together and with a third tube down which we can blow. Now suppose that we blow soap bubbles on the ends of the tubes. We start with a flat film across the openings; as we blow these two films swell up symmetrically until they are both equal hemispheres on the ends of the tubes. If we then blow a bit more a strange thing happens: instead of two bubbles getting bigger and bigger as we might imagine, one bubble indeed gets bigger but the other collapses back and gets smaller. Three stages are shown below.

This can also be explained in terms of minimization of potential energy. The force of gravity encourages objects to seek the lowest point, whereas in the case of soap films, the system seeks to become that with minimum surface area. The one large bubble has a smaller surface area than two smaller bubbles with the same total volume.

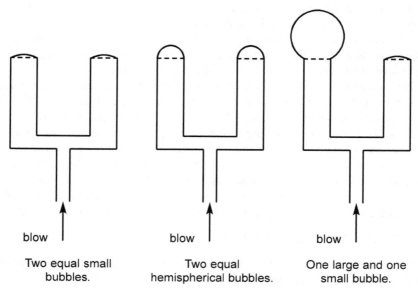

Two equal small bubbles.   Two equal hemispherical bubbles.   One large and one small bubble.

Two bubbles

A SINGLE BUBBLE naturally forms a sphere. But let us consider what shapes we get when two bubbles meet. Three films are formed which are the remaining parts of the original two bubbles plus a film between them. These three films meet together in a circle. Each film is of course part of a sphere except that if the original two bubbles are exactly the same size then the joining film will be flat.

Consider the forces on a point of the circle where the three films meet. We know that the surface tension is the same throughout and therefore at that point there will be three equal forces acting on the point. It follows by symmetry that the three films must be at the same angle to each other and since the whole angle at a point is 360°, the angle between the three films must be 120°.

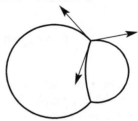

This is just a special case of Lami's theorem in statics named after the French mathematician Bernard Lami (1640–1715). It states that if three forces act at a point in equilibrium then the angles must be such that each force is proportional to the sine of the angle between the other two forces. This is really the sine theorem which we met in Lecture 4 when discussing cross-ratios.

Now suppose that one bubble has radius R_1 and the other bubble has radius R_2 and that the radius of the intermediate film is R_3 as in the diagram below. The centres of the two bubbles are O_1 and O_2 and the centre of curvature of the intermediate film is O_3.

Since the tangents to the three films are at 120° to each other it follows that the radii are also at 120° to each other. As a consequence the angles O_1PO_2 and O_2PO_3 are both 60°. PX is the tangent to the intermediate film and thus at right angles to PO_3 from which it follows that O_1PX and XPO_2 are both 30° so that PX bisects O_1PO_2 as shown below.

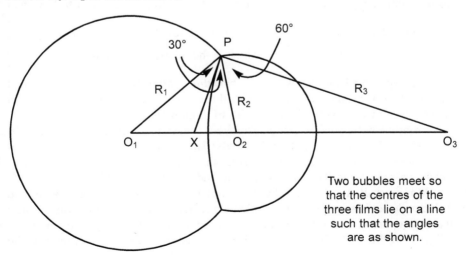

Two bubbles meet so that the centres of the three films lie on a line such that the angles are as shown.

140   Gems of Geometry

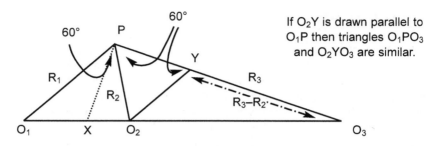

Proving that $1/R_3 = 1/R_2 - 1/R_1$ using similar triangles.

The radius of the intermediate film R_3 is related to the radii of the two bubbles by a simple formula. Recall that the pressure in a bubble is inversely proportional to its radius since $P = 4T/R$. The radius of the intermediate film is dictated by the difference in the pressures on either side of it. These pressures are $4T/R_1$ and $4T/R_2$ respectively. It immediately follows that $P_3 = 4T/R_3 = 4T/R_2 - 4T/R_1$. So finally we have the simple equation

$1/R_3 = 1/R_2 - 1/R_1$ relationship between the radii of the three films

This relationship can also be deduced from the fact that the three films are at 120° to each other. In the diagram above we have omitted the films for simplicity and the tangent line is shown dotted. Now if we draw the line O_2Y so that it is parallel to the line O_1P then the angle PO_2Y is the same as O_1PO_2 from which it follows that all the angles of the triangle PYO_2 are 60° so that it is indeed equilateral with all sides being of length R_2. This means that the sides YO_2 and YO_3 of the triangle O_2YO_3 are R_2 and R_3-R_2 respectively.

Now the triangle O_2YO_3 is similar to the triangle O_1PO_3 having corresponding angles the same and so the sides are in proportion. It follows that

$(R_3-R_2)/R_2 = R_3/R_1$ if we now divide throughout by R_3 we get

$1/R_2 - 1/R_3 = 1/R_1$ which is the relationship we want.

(It is perhaps surprising that we can deduce this relationship simply from the fact that the three films meet at 120°.) We will use the argument of this discussion in reverse when we consider the properties of three bubbles.

From the same similar triangles above it is easy to see that the ratio of O_1O_3 to O_2O_3 is the same as that of R_1 to R_2. This is a neat result: the centre of curvature of the intermediate film divides the line joining the centres of the two bubbles externally in the same ratio as the radii of the two bubbles.

Moreover, it is a standard theorem of elementary geometry (see note at end of lecture) that if an angle of a triangle be bisected then the bisecting line divides the side opposite the angle in the same proportion as the lengths of the sides adjacent to the angle. So since PX bisects the angle O_1PO_2 this means that the

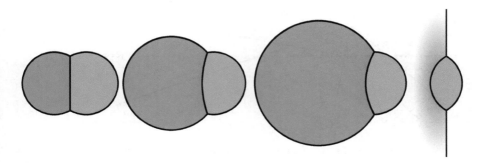

A sequence of bubble pairs where the left grows so large that it eventually becomes a flat film.

ratio of O_1X to O_2X is the same as that of R_1 to R_2. So the tangent divides the line joining the centres of the two bubbles internally also in the same ratio as the radii of the two bubbles.

Putting these two results together we have the pretty result that O_1, O_2, O_3 and X form a harmonic range as discussed in the lecture on projective geometry when we were looking at cross-ratios. Here is harmony in the spheres although perhaps not quite what Pythagoras had in mind!

It is interesting to consider some typical values of the radii. In the diagrams above the first shows two bubbles of radius 2 units with a flat intermediate film.

In the second the left bubble has a radius of 3 units and the right has a radius of 2 units. It then follows that $1/R_3 = 1/2 - 1/3 = 1/6$ so that the intermediate film has a radius of 6 units. The distance O_1O_3 can be shown to be $3\sqrt{7}$ units and O_2O_3 is $2\sqrt{7}$ units so that the ratio of these distances is also 3 to 2 as expected.

In the third case the left bubble has grown to a radius of 4 units and is twice that of the right bubble. The radius of the intermediate film is then the same as that of the left bubble since $1/2 - 1/4 = 1/4$ and, moreover, the surface of the right bubble is exactly a hemisphere (the triangle O_1PO_2 is half the isosceles triangle O_1PO_3 and so has a right angle at O_2). The distance O_1O_3 is $4\sqrt{3}$ whereas O_2O_3 is $2\sqrt{3}$ which are again in the same ratio as the radii of the bubbles.

In the final case the left bubble has become so large that it is a flat film. It then follows that R_3 is the same as R_2 and the result is a lenticular bubble hanging in a flat film.

We conclude by returning to the pencil of four lines PO_1, PX, PO_2 and PO_3. We recall from the discussion on cross-ratio in the lecture on projective geometry that such a pencil cuts any line in four points in the same cross-ratio. Indeed we showed that the angles between the lines are 30°, 30°, and 60° without considering the size of the bubbles. It follows therefore that *any* line cutting these four lines will cut them in points which form a harmonic range and so correspond to the centres of a pair of bubbles, the centre of curvature of the intermediate film and its tangent intersection point.

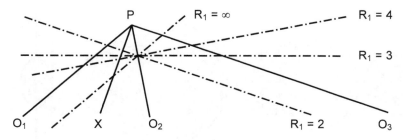

Lines of centres for pairs of bubbles with $R_2 = 2$.

The diagram above shows the pencil plus the lines of centres corresponding to the sequence of bubbles as dashed lines. Two special cases arise. When the line of centres is parallel to PO_3, the intermediate film is flat which corresponds to equal bubbles and when the line of centres is parallel to PO_1 or PO_2, then we have the case of the lenticular bubble hanging in a film.

We can go one stage further by considering the tangents to the bubbles themselves at the point P. Suppose these meet the line of centres in points X_1 and X_2 respectively. This is shown below although X_2 has unfortunately gone off the page to the left! Note that all the angles at P are 30°. By similar arguments to before we find that the points X_1 and X_2 also form harmonic ranges with the three centres taken in an appropriate order. If we relabel X as X_3 then we find that

X_1 and O_1 divide O_2O_3 internally and externally in the ratio of R_2 to R_3;

X_2 and O_2 divide O_3O_1 internally and externally in the ratio of R_3 to R_1;

X_3 and O_3 divide O_1O_2 internally and externally in the ratio of R_1 to R_2.

It is amazing that there is so much geometric beauty in just two soap bubbles.

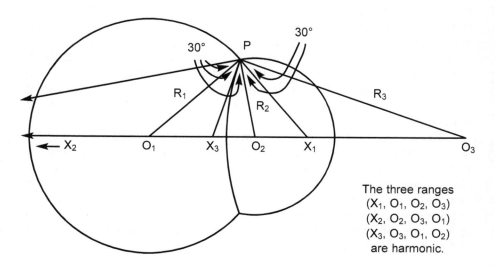

The three ranges
(X_1, O_1, O_2, O_3)
(X_2, O_2, O_3, O_1)
(X_3, O_3, O_1, O_2)
are harmonic.

Three bubbles

NOW SUPPOSE that a third bubble is placed so that it touches the two bubbles we have just been discussing. Because we can always draw a plane through any three points it follows that the three bubbles are symmetric about the plane through their centres. The result is as shown below.

For simplicity we have changed the notation slightly. The centres of the three bubbles are now O_1, O_2 and O_3 with radii R_1, R_2 and R_3. The centre of curvature of the film between bubbles 1 and 2 is the point O_{12} and the radius of curvature is R_{12}. In a similar way we denote the centres of curvature of the films between bubbles 1 and 3 and between bubbles 2 and 3 as the points O_{13} and O_{23} with radii of curvature R_{13} and R_{23}.

We know of course that O_1, O_2 and O_{12} lie on a straight line because adding the third bubble does not disturb the configuration of two bubbles around the point P provided that the pressures remain the same. Similarly O_1, O_3 and O_{13} lie on a line as do O_2, O_3 and O_{23}. What is perhaps rather amazing is that the centres of the three intermediate films O_{12}, O_{13} and O_{23} also lie on a straight line.

As consequence, the six points lie in groups of threes on four lines. The four lines are the four lines of a complete quadrilateral (in blue) and the six points are the six points where the four lines meet. Note that each line corresponds to an intersection of three films.

To prove this, we use a similar argument to that of the previous section but in reverse. Suppose the three films meet at the point P as shown overleaf.

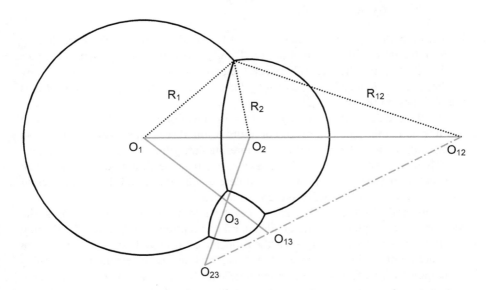

The centres of curvature of the six films formed by three bubbles meeting form the six points of a complete quadrilateral.

144   Gems of Geometry

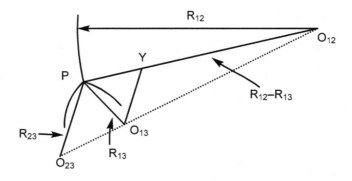

Proving that O_{13} lies on the line joining O_{23} and O_{12}.

First draw a line parallel to $O_{23}P$ through O_{13} meeting PO_{12} at Y; the triangle $O_{13}YP$ is then equilateral as before because the films meet at 120°. It then follows that the length of YO_{12} is the length of PO_{12} (which is R_{12}) minus that of PY (which is the same as PO_{13} which is R_{13}). So the length of YO_{12} is $R_{12} - R_{13}$.

If the pressures in the three bubbles are P_1, P_2 and P_3 then since the radii of the films are dictated by the differences in the pressures we have $P_2 - P_1 = 4T/R_{12}$ and so on from which it follows that

$$1/R_{12} + 1/R_{23} = 1/R_{13}$$ and then by multiplying by R_{12} we get

$$(R_{12} - R_{13})/R_{13} = R_{12}/R_{23}$$

It then immediately follows that the triangles $O_{23}PO_{12}$ and $O_{13}YO_{12}$ are similar because the ratio of the pair of sides $O_{23}P$ and PO_{12} is the same as that of the pair $O_{13}Y$ and YO_{12} and the angles between the pairs of sides are the same since they are both 120°. Hence the angle $O_{23}O_{12}P$ is the same as $O_{13}O_{12}Y$. It then follows that O_{13} lies on the line $O_{23}O_{12}$.

So we see that corresponding to every intersection between three films there is a line containing the centres of curvature of the three films. Well that is not so surprising, the joint between the three intermediate films is just like any other.

Thus we have now proved the nice result that the four lines, one for each intersection between films, form a complete quadrilateral and the six centres of curvature are the six points of the quadrilateral.

Now add the diagonal triangle LMN (in red, see opposite) joining the opposite pairs of points of the quadrilateral as mentioned in Lecture 4. We may recall that the diagonal triangle has the property of forming harmonic ranges with the other points and lines. As a consequence, the line NO_3 meets the line $O_1O_2O_{12}$ in the point which forms a harmonic range with the three points O_1, O_2 and O_{12}. However, we saw in the last section that the point X where the tangent to the intermediate film between bubbles 1 and 2 meets the line of centres is such that O_1, O_2, O_{12} and X form a harmonic range.

So we have discovered that the line NO_3 actually goes through the point X. Similar results follow for the other tangents at P and so we deduce that the line $O_{23}L$ goes through X_1 and $O_{13}M$ goes through X_2 which is off the page. Similarly, it can be shown that the line NO_3 meets the line $O_{23}O_{13}O_{12}$ in the point where the tangent to the intermediate film between bubbles 1 and 2 at the meeting point between the three intermediate films cuts that line.

Altogether there are twelve points where tangents meet the lines of centres – three points on each of the four lines of centres. The six lines LO_1, MO_2, NO_3 and $O_{23}L$, $O_{13}M$, $O_{12}N$ each go through two of these twelve points.

The figure formed by the three bubbles is intriguing and abounds with harmonic ranges revealing further harmony between the spheres of the bubbles.

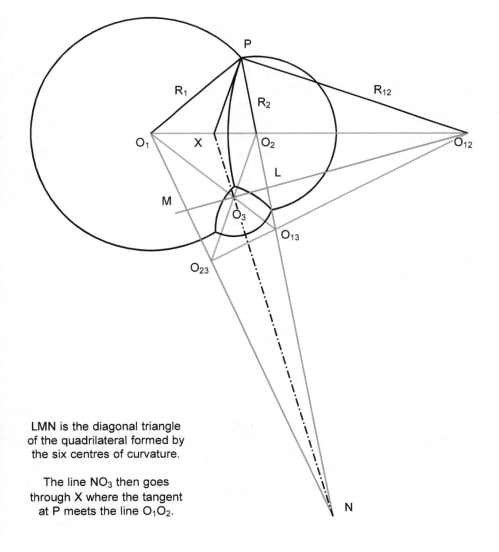

LMN is the diagonal triangle of the quadrilateral formed by the six centres of curvature.

The line NO_3 then goes through X where the tangent at P meets the line O_1O_2.

Four bubbles

NOW ADD A FOURTH BUBBLE so that it meets the three bubbles. The centres of the four bubbles no longer lie in a plane and so it is hard to draw a satisfactory diagram. However, each group of three bubbles acts as we have just described and their centres lie in a plane containing six centres of curvature.

There are ten films, four are the bubbles themselves and, since there are six ways of choosing two bubbles from four, there are six intermediate films making ten altogether. We can denote the centres of the four bubbles by points O_1, O_2, O_3, O_4 and the centres of the intermediate films by O_{12}, O_{13}, O_{14}, O_{23}, O_{24}, O_{34}. Also, we have just seen from studying three bubbles that these points lie on lines in groups of three.

This should bring another echo of the lecture on projective geometry. In fact, the ten points are arranged as the points of the Desargues configuration which we saw lie on ten lines in groups of three. The configuration is shown below with the points named according to the convention just described. (The points for the case of three bubbles are at the bottom of the diagram and are arranged much as in the previous section except that they are reflected left for right.)

The lines have been named as well. The line joining the centres of bubbles 1 and 2 is l_{12} and so on whereas the line joining the centres of curvature of three intermediate films is numbered according to the bubble not involved; thus the line of the centres O_{12}, O_{13} and O_{23} is l_4.

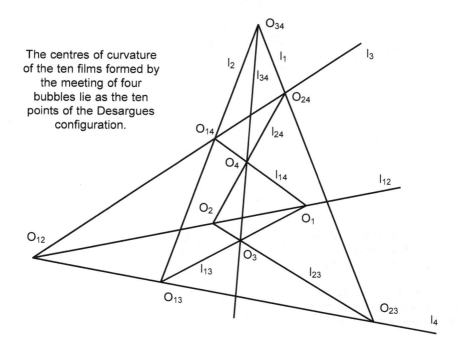

The centres of curvature of the ten films formed by the meeting of four bubbles lie as the ten points of the Desargues configuration.

This diagram reveals an intriguing fact. We expect there to be four planes because clearly the centres of the bubbles define four planes. However, the Desargues configuration contains five planes so where does the other plane come from? The answer is that the centres of curvature of the six intermediate films all lie in the same plane and this is the fifth plane. The six centres and the four lines l_1, l_2, l_3 and l_4 form a complete quadrilateral in this plane.

I find it rather amazing that a group of humble bubbles should relate to one of the most fundamental configurations of projective geometry. But that is a typical illustration of the beauty of geometry. Seemingly unrelated topics suddenly have a deep resonance. Well on this high note of ecstasy we will leave the four bubbles and in a sense the story stops there because we have run out of dimensions – we cannot add a fifth bubble so that all five bubbles touch each other (at least not in this world).

Foam

WE WILL NOW turn to a consideration of how lots of bubbles arrange themselves. This is a rather complex problem and so we will content ourselves with just looking at two simple cases.

Suppose we have two sheets of glass and some bubbles trapped between them to form a single layer. We will also suppose that the bubbles wet the glass so that they are at right angles to it. The bubbles will of course meet each other at 120° as usual. Moreover, if the bubbles all have the same size and are at the same pressure then the films between them will be flat and the result will be a uniform mass of regular hexagonal prisms much like a honeycomb as shown below.

It is interesting to consider the shape of the bubbles at the edge of the honeycomb (the diagram below shows a section of honeycomb without the edge). For simplicity suppose we have seven bubbles with one bubble in the centre surrounded by the other six. Then the outside surfaces will be curved but

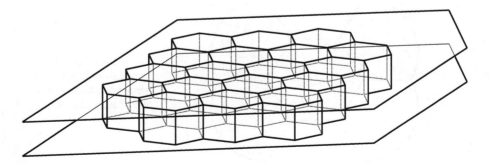

Portion of hexagonal honeycomb of bubbles between two pieces of glass.

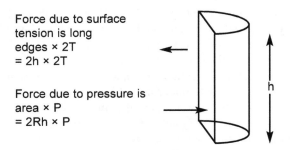

Force due to surface tension is long edges × 2T = 2h × 2T

Force due to pressure is area × P = 2Rh × P

The forces on half a cylindrical bubble.

since we have assumed that the bubbles wet the glass the curved surfaces will be cylindrical rather than spherical.

As a brief detour let us do the calculation of the pressure difference across a cylinder from first principles as we did for the sphere at the beginning of this lecture. For a section of half cylinder of length h and radius R we see from the diagram above that

$4hT = 2RhP$ and so $P = 2T/R$ pressure in cylinder

So the pressure in a cylinder is one half of that in a sphere of the same radius. This in fact is just a special case of a general formula. A general surface has two radii of curvature (these correspond to the directions of greatest and least curvature) and if these are R_1 and R_2 then the pressure formula is

$P = 2T/R_1 + 2T/R_2$ pressure across general surface

In the case of a sphere R_1 and R_2 are the same thus giving $4T/R$ whereas in the case of the cylinder one radius is infinite so one term vanishes leaving $2T/R$.

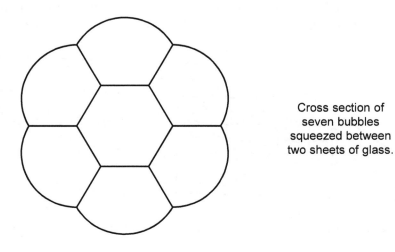

Cross section of seven bubbles squeezed between two sheets of glass.

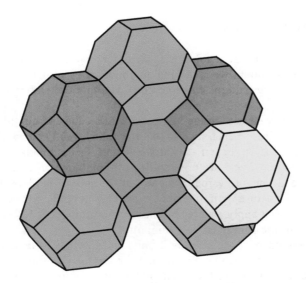

In three dimensions a possible ideal foam approximates to a honeycomb of truncated octahedra.

And now to return to the seven bubbles. If we suppose that the pressure in each is the same then the resulting shapes will be as shown opposite. The outer bubbles have three flat faces and one cylindrical one. If the flat faces are all the same size, the surrounding bubbles have a smaller volume than the central one.

Let us now consider what happens in three dimensions. Again we will suppose that all the bubbles are of the same size and more importantly have the same pressure which ensures that all intermediate faces have no curvature. It turns out there are always three bubbles meeting at each line between them and therefore that the faces have to be at 120°. The only regular honeycomb that fits the bill is that of rhombic dodecahedra, one of the dual Archimedean figures which we met in Lecture 3 when we were about to embark into Four Dimensions. So we might expect a foam of equal sized bubbles to be composed of lots of rhombic dodecahedra. But this is not the case. Although the rhombic dodecahedron does indeed have its faces at 120° and also has the least surface area of all the honeycombs for the volume contained, nevertheless it is not the correct answer. One problem with this honeycomb is that a rhombic dodecahedron is not that regular, some vertices have four lines meeting and others have six.

Another possibility is the honeycomb of truncated octahedra which also has three faces on each edge and moreover is regular in that every vertex of a truncated octahedron is the same and has just three lines meeting. But the problem here is that the three angles where the faces meet are not 120°. One (that between the hexagonal faces) is the familiar angle of 109° 28' and the other two are 125° 16'. The answer will become clear in a moment.

Films on frames

IF WE DIP a wire loop into a solution of detergent and remove it we will find a film across it. No doubt we recall making bubbles by blowing through such a loop. Some intriguing patterns can emerge if we dip more elaborate shapes into a solution.

Suppose we have a wire frame in the shape of a cube and dip this. One might expect to obtain a film on each face so that the cube is really constraining a large bubble. But this doesn't happen – what we actually get is much like that shown in the diagram below. There are thirteen films in total, one on each of the twelve edges and these all meet towards the centre of the cube around a square looking film. This is perhaps disappointing since it is not symmetric. However, there are two general laws that apply, one is that the films must meet at 120° and the other is that the area of the film must be a minimum.

Suppose the edge of the cube is 2 units and that the little square film has edge $2x$ units. If we assume that all the films are flat then it turns out that the least area occurs when x is about 0.073 units. This is a tedious but straightforward calculation; there are 8 trapezoidal films, 4 triangular films and one square one. It is easy enough to compute the area of all these and then find the condition for the area to be a minimum. This gives the value of x mentioned.

On the other hand if we consider the fact that the films meet at 120° then by considering the meeting of the trapezoidal films with the square film we can deduce that x must be 0.423 which contradicts our previous value.

Obviously we have made a wrong assumption somewhere. If we make a wire cube and dip it then all becomes clear. The square in the centre is not a square at all but has slightly curved sides. Moreover, the trapezoidal films are not flat either. Recall that the general formula for the pressure difference across a film is

$$P = 2T/R_1 + 2T/R_2 \qquad \text{pressure across general surface}$$

So if the film is shaped like a saddle with equal but opposite curvatures in two directions at right angles then the pressure difference will be zero but the film

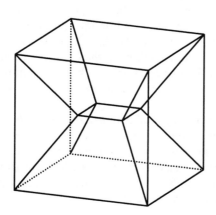

Approximate shape of film on wire cube.

There are 8 trapezoidal films, 4 triangular films and one square film.

In fact the square film has slightly convex sides and so the trapezoidal films are not quite flat.

will not be flat. And this indeed is what happens. The trapezoidal films are ever so slightly saddle shaped. The triangular films are indeed flat by symmetry and so of course is the curvilinear square in the middle.

Now compare the honeycomb of truncated octahedra with the films at the centre of the wire cube and in particular consider the truncated octahedra surrounding one square face. It is immediately clear that the squarish film corresponds to the square face, the trapezoidal surfaces correspond to parts of the hexagonal faces adjacent to the square face and the triangular faces correspond to parts of the adjacent square faces which are at an angle to the central square face. So the puzzle is solved.

Our ideal regular foam thus consists of slightly distorted truncated octahedra whose hexagonal faces are not quite flat and whose square faces (although flat) have slightly curved edges. These curvatures ensure that the faces meet at 120° as required. Moreover, this distorted figure has slightly less surface area per volume than the rhombic dodecahedron and so wins on that score as well. A truncated octahedron has 14 faces and so each bubble has 14 neighbours.

Very recently, a slightly more efficient arrangement in terms of surface per volume was discovered by Denis Weaire and David Phelan of Trinity College, Dublin. This consists of a mixture of 12-gons and 14-gons of the same volume. The 12-gons are irregular dodecahedra with irregular pentagonal faces whereas the 14-gons are also irregular and have 12 pentagonal and two hexagonal faces. In practice it is extremely difficult to create a foam whose bubbles all have exactly the same volume so these arrangements are unlikely in a real foam. Typically a foam is made of bubbles of somewhat varying size and the number of faces varies from perhaps 12 to 16.

Films on cylinders

THAT A FILM may be curved and yet flat (which sounds like a contradiction) is perhaps best demonstrated by considering a soap film across two rings.

Suppose we capture a bubble between two rings and then move the rings so that the part of the film between the rings is a cylinder. Then we will have two large parts of spheres on the rings as shown below. If the radius of the cylinder

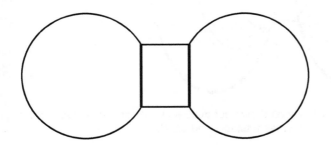

Cross-section through two bubbles and a cylinder of film held on two rings.

is R_c, then the pressure in the cylinder is $2T/R_c$; if the radius of the spheres is R_s then the pressures are $4T/R_s$. Since these pressures must be the same, it follows that the radius of the two big bubbles must be exactly twice that of the cylinder.

If we now break one of the bubbles (the rings must not be too far apart) then the other will become flat whereas the film between the rings will take a waisted form as shown below. This portion of film is of course open to the air and so its total curvature must be zero. We conclude that the curvature everywhere must be equal and opposite in the two principal directions.

The shape of the cross-section is known as a catenary and is the same shape as that adopted by a hanging chain. (The Latin for chain is *catena*, whence the name.) The solid figure obtained by rotating a catenary is usually called a catenoid.

The catenary has some interesting properties. An important one relates to the fact that the catenoid has zero total curvature. The radius of curvature at any point on a catenary is exactly equal to the distance between that point and where the normal meets the axis. The radius of curvature is that of the circle that most closely fits the curve at the point and the normal is at right angles to the tangent. So in the diagram opposite *CP* equals *PN*.

As a special case the radius of curvature at the centre of the catenary equals the distance from the catenary to the axis which we can suppose has length *a* as shown in the inverted catenary at the bottom of the diagram opposite.

It should now be clear why the catenoid has zero total curvature. The curvature in the plane of the paper at *P* has radius *CP*. On the other hand the curvature in the plane which is at right angles to the paper (and which of course

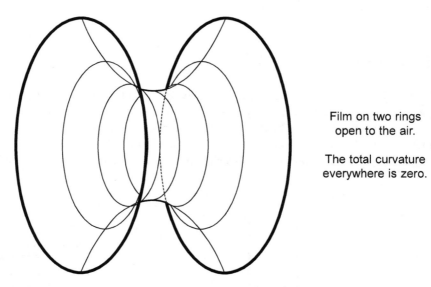

Film on two rings open to the air.

The total curvature everywhere is zero.

P S. The film cannot actually be drawn out quite so far as shown above since it becomes unstable and breaks.

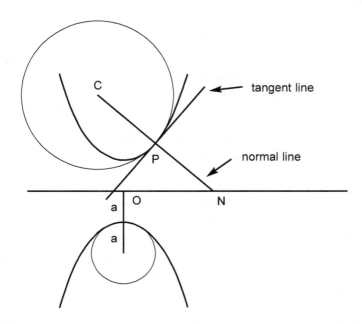

Cross-section through a catenoid showing catenaries and circles of curvature. The distance CP is equal to PN.

contains the normal line) has radius *PN*. But as just mentioned it is a property of the catenary that these are equal and so the two curvatures cancel out.

If we now slowly pull the two rings apart the waist of the catenoid narrows; one might expect that the waist would eventually contract to nothing but before this happens the film always breaks. In fact if the rings have radius *R* then the distance between the rings for a stable film cannot exceed about 1.325*R*. The minimum waist (2*a*) is about 1.1*R*. Any attempt to pass these limits results in an unstable film which breaks. This corresponds to when the tangent to the point *P* goes through the central point *O*.

For those interested in the detailed mathematics, the equation of a catenary is $y = a \cosh(x/a)$ where the cosh function is the hyperbolic cosine usually defined as $\cosh x = (e^x + e^{-x})/2$ where *e* is the base of natural logarithms.

Another example of instability is illustrated by considering a cylindrical film. If we take the case of two rings with bubbles on the ends and a film between the rings and somehow contrive to pull the rings apart and at the same time pump more air in so that the cylindrical film remains a cylinder then eventually the cylinder becomes unstable, goes wobbly and breaks up into various bubbles. This instability happens when the length of the cylinder is exactly equal to its circumference. In other words when the length is 2π*R*. Incidentally, the Greek letter π is used to denote the ratio of circumference to diameter because π is the first letter of the Greek word περιμετρος (*perimitros*), perimeter.

That well known theorem

ON PAGE 140, the proof of the fact that O_1, O_2, O_3 and X form a harmonic range used the hopefully well known theorem that if an angle of a triangle be bisected then the bisecting line divides the side opposite the angle in the same proportion as the lengths of the sides adjacent to the angle. This is fairly easy to prove.

Consider the triangle ABC and suppose the line AX bisects the angle at A and meets BC at X. We have to show that $AB/AC = BX/CX$.

Drop perpendiculars BP and CQ from B and C onto the line AX. Now the right-angled triangles BPX and CQX are similar because they have the same angle at X. Hence their sides are in the same ratios and therefore

$BX/CX = BP/CQ$

Now the right-angled triangles BPA and CQA are also similar because they have the same angle at A. Therefore

$BP/CQ = AB/AC$

If we put these two relationships together we get

$BX/CX = AB/AC$

which is what we wanted to prove. QED

Well that was really very easy. But the curious thing is that it seems to be ignored by most textbooks. It is not one of Euclid's propositions.

Further reading

THE OBVIOUS starting point for further reading is *Soap Bubbles* by Boys; this is a delightful book and covers many aspects of bubbles as well as other aquatic forms such as jets of water. It has some very pretty drawings. However, Boys doesn't mention that the squarish film at the centre of the cube is in fact curved. A brief discussion of the discovery by Weaire and Phelan of the more efficient arrangement with a mixture of irregular figures will be found in *Kepler's Conjecture* by Szpiro.

Another interesting discussion will be found in *On Growth and Form* by D'Arcy Thompson; this book generally is about the shape of living organisms.

Regarding bubbles, he relates the shapes of bubbles to that of cells in organisms and the stability of cylinders to why drops form on a spider's web and so on. A rather different discussion from the point of view of minimization of the area of a surface will be found in *What is Mathematics?* by Courant and Robbins.

Both D'Arcy Thomson and Courant and Robbins have many illustrations suggesting other experiments with films. It is interesting, for example, to consider the films obtained when figures such as a tetrahedron, octahedron and triangular prism are dipped.

Exercises

1 Two soap bubbles are joined. One has a radius of 1 cm and the other has a radius of 3 cm (after they are joined together). What is the radius of the intermediate film between them?

2** Seven bubbles are squeezed between two parallel sheets of glass such that the intermediate films are all flat and the same size as shown below. The volume of the central bubble is 1 cc. What are the volumes of the surrounding bubbles?

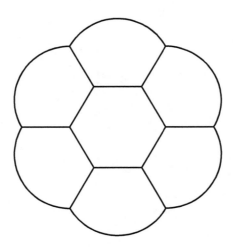

7 Harmony of the Spheres

IN THE LECTURE on Bubbles we saw how the centres of bubbles, the centres of curvature of dividing films, and points on tangents formed various harmonic ranges. In this lecture we will start by looking at some other pretty properties of circles and spheres. But first a brief explanation of the title of this lecture.

It seems that the phrase "music of the spheres" originated from Pythagoras and it was really all about the motion of the planets. Plato said that a siren sits on each planet who carols a most sweet song, agreeing to the motion of her own particular planet, harmonizing with others. Hence Milton speaks of the "celestial syrens' harmony that sit upon the nine enfolded spheres". Maximus Tyrius says that the mere proper motion of the planets must create sounds and as the planets move at regular intervals, the sounds must harmonize.

There are a number of properties of circles and spheres which are simply stated but surprisingly tricky to prove unless we get the right view of the problem. So this lecture is really about how the appropriate transformation can turn a hard problem into an easy one. In a later lecture we will look at how the complex plane can easily reveal a few unexpected properties of figures involving polygons.

Steiner's porism

THEOREMS come in various categories and only the best seem to deserve the term theorem. I suppose a Theorem is a very important fact from which other things can be proved. Pythagoras theorem is an obvious example of a jolly good theorem.

A subsidiary conclusion leading on immediately from a theorem is often called a Corollary. This comes from the Latin *corollarium* meaning originally a garland of flowers such as given to an actor and then later meaning a gratuity or free gift. So a corollary comes for free from a theorem with no further effort.

A preliminary conclusion which is needed in order to further the proof of a theorem is called a Lemma. This comes from the Greek λεμμα (*lemma*), meaning a skin or base and thus something assumed. Being of Greek origin the posh plural of lemma is of course lemmata.

And finally there is a Porism which seems to be a second class theorem which leads nowhere. Porism comes from the Greek ποριζω (*porizo*), which roughly means to find or obtain. Anyway, a curious discovery of the Swiss mathematician Jakob Steiner (1796–1863) is traditionally known as a porism.

Draw two circles one inside the other and with different centres and thus not concentric. Now draw a circle touching both original circles; then draw a circle

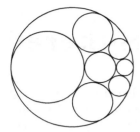
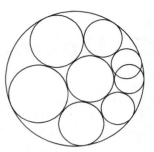

Steiner's porism.

If the chain of circles joins up (left) then it always does wherever the chain starts.

If not (right) then it never does.

touching this circle and both original circles. Continue in this way and so draw a series of circles touching the original circles and each other to form a chain. It might be that the last circle just touches the first so that the chain of circles is neatly closed as in the left-hand example above. More often than not this does not happen as in the other example. But the amazing fact is that whether the circles form a closed chain or not does not depend upon the position of the first circle of the chain.

To prove this using brute force would be very nasty but it can be shown to be almost obvious using an ingenious transformation known as inversion.

Inversion

A TRANSFORMATION is essentially a rule for converting one configuration (such as a diagram of lines, circles, etc) into a related one. Certain properties of the original configuration will remain preserved but others might not be. If we can prove something in the transformed version and the necessary properties are preserved then the corresponding thing will be true in the original.

A very simple transformation is that obtained by reflection in a flat mirror. This is unhelpful because the new configuration is pretty much the same. For example, the reflection of an equilateral triangle will still be an equilateral triangle. So anything hard to prove in the real world will remain just as hard to prove in the Looking Glass world – although in Alice's case the Looking Glass world didn't seem to be quite the same at all!

The important transformation known as inversion is a sort of reflection in a circle. Suppose the circle has centre O and radius R as shown opposite; we call O the centre of inversion. Then the rule is that any point P is transformed into a point P' such that P' is on the line OP and the distance OP multiplied by OP' is equal to the square of the radius R. Thus

$$OP \times OP' = R^2 \quad \text{fundamental rule for inversion}$$

Inversion goes in reverse obviously; thus if P transforms into P' then P' transforms into P. (Unprimed points and lines are in red and the primed inverses

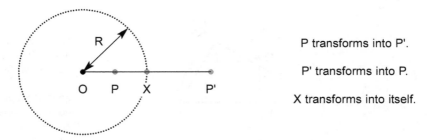

The basic rules of inversion.

are in blue.) Note that every point inside the circle transforms into a point outside the circle and vice versa. The only points that transform into themselves are the points on the circle of inversion such as X (which is shown red and blue). So the circle of inversion as a whole transforms into itself.

It is clear that every point of a line through the centre of inversion O transforms into a point of the same line. So the line as a whole transforms into itself. But the individual points on the line do not transform into themselves except for the point X where the line meets the circle of inversion. It's best not to ask what the point O transforms into (we cannot consider it to be a line at infinity because that breaks the otherwise general rule that a point transforms into a point).

We will now consider the transformation of lines not through O. The diagram below shows the case where the line PQ does not meet the circle of inversion. P is the point on the line such that OP is at right angles to the line and Q is any other point on the line. P' and Q' are the inverted points. Now consider the two triangles OPQ and $OQ'P'$. By the rule for inversion we know that $OP \times OP' = R^2 = OQ \times OQ'$ from which it follows that the two triangles are similar and therefore that the angle $OQ'P'$ is also a right angle. It follows that Q' lies on the circle whose diameter is OP' since the angle in a semicircle is a right angle

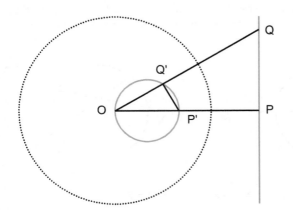

The inverse of the line PQ is the circle OP'Q'.

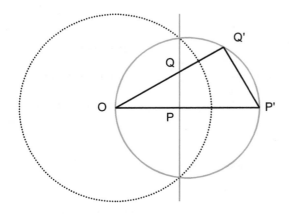

The inverse of the line PQ is the circle OP'Q'.

It goes through the two points where the line PQ meets the circle of inversion.

(Euclid Book III, Proposition 31). So finally we deduce that the inverse of the line *PQ* is in fact the circle *OP'Q'*. The same argument applies if the line meets the circle of inversion as shown above.

So the inverse of a general line is a circle through the centre of inversion *O*. Conversely, the inverse of a circle through the centre of inversion is a straight line not through the centre of inversion. Note that if the line intersects the circle of inversion then the inverse circle also goes through the same two points on the circle of inversion.

Finally, we have to consider the case of a circle not through *O*. The key to this is a couple of Euclid's theorems (Book III, Propositions 35 and 36) which are illustrated below. *P* is any point outside a circle and *PT* is a tangent to the circle. *PXX'* is a line which meets the circle in the points *X* and *X'*. Then the product of *PX* and *PX'* equals the square of *PT*. Also if two lines through *P* meet the circle in *X*, *X'* and *Y*, *Y'* respectively then the product of *PX* and *PX'* equals the product of *PY* and *PY'*. This applies whether *P* be inside or outside the circle.

Now in the diagram opposite, suppose we start with circle *XYT* with centre *C*. Note that *XY* is any line through the centre of inversion. Suppose that the product of *OX* and *OY* has the value *p*; then by Euclid's theorem just mentioned this value *p* is the same no matter what line we take through *O*. Now draw the circle which is the factor R^2/p larger than the original circle and whose centre *D*

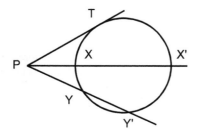 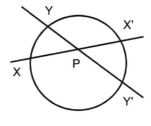

Euclid III, 35,36:   PX × PX' = PY × PY' = PT²

7 Harmony of the Spheres

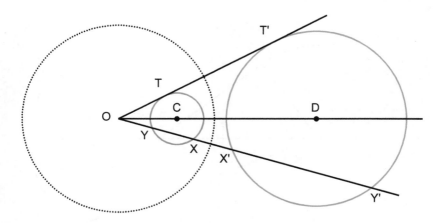

Circle X'Y'T' is the inverse of circle XYT. Case where O is outside the circles.

is the same factor further from O. Remember that R is the radius of the circle of inversion. Suppose this new circle meets the line in X' and Y' as shown. Then

$OX' = OY \times R^2/p$ because of expansion factor, and
$OX = p \div OY$ because of definition of p, so multiplying together
$OX \times OX' = R^2$ and so X' is the inverse of X.

This argument follows for any point X on the original circle and so the new circle is indeed the required inverse of the original one. And vice versa of course. Note however, that the centre of the new circle D is *not* the inverse point of the centre C of the original circle (which is why we didn't call it C'). The diagram above shows the case where the centre of inversion is outside the two circles; it also shows the tangent line OTT' where T' is the inverse of T.

A similar argument applies if the centre of inversion is inside the two circles. However, in this case we cannot show the tangents. Other possibilities are where the circles intersect the circle of inversion. Both the original and inverse circles meet the circle of inversion in the same two points. These cases are shown below.

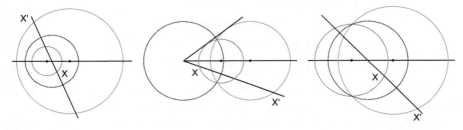

Centre of inversion inside.   Cases where circles intersect the circle of inversion.

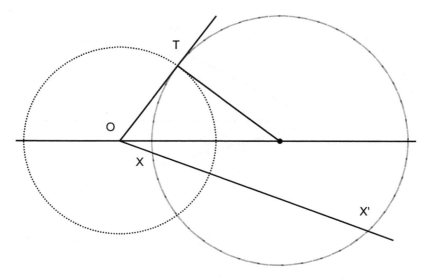

A circle meeting the circle of inversion at right angles inverts into itself.

A very important situation is shown above. This is when the original circle meets the circle of inversion at right angles (that is the two circles are orthogonal to each other so that the tangent of each goes through the centre of the other). The tangent OT to the circle then intersects the circle of inversion at T so that T inverts into itself. In this special case the whole circle inverts into itself. That this is so follows immediately from Euclid's theorem mentioned above since by that theorem $OX \times OX' = OT^2$ but this is precisely the rule for inversion. Note carefully that the individual points on the circle do not invert into themselves; the arc inside the circle of inversion transforms into the arc outside.

There is however one circle that does invert into itself point for point and that of course is the circle of inversion.

We have at last covered all the rules for the inversion of lines and circles which can be summarized thus (O is the centre of the circle of inversion, C).

line through O <=> same line through O,

line not through O <=> circle through O,

circle not through O <=> circle not through O,

circle orthogonal to C <=> same circle orthogonal to C,

circle of inversion, C <=> itself, point for point.

We will now briefly consider what properties are preserved by inversion. It is obvious that lengths are not preserved but perhaps surprising that angles are always preserved. Thus if two lines intersect at right angles then the circles which are their inverses will be orthogonal to each other.

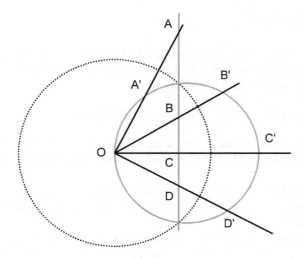

The four points A, B, C, D on the line have the same cross-ratio as the four points A', B', C', D' on the circle.

(The proof that angles are preserved is left for the reader; draw a short curve PQ and its inverse $P'Q'$, then triangles OPQ and $OQ'P'$ are similar using the inversion formula and so the angles OPQ and $OQ'P'$ are equal; then let Q approach P so PQ becomes the tangent. This shows that the angle between any line and a radial line is preserved; the general case follows by adding or subtracting angles.)

Although lengths are not preserved, it will perhaps not come as a surprise that cross-ratios are preserved. This is very easy to show for any four points on a line through O by using the formula for cross-ratio and the rule that $OX \times OX' = R^2$.

In the case of a line not through O, four points A, B, C, D on the line invert into four points A', B', C', D' on a circle as shown above. This is a good moment to define the cross-ratio of four points on a circle. Take another point P on the circle and consider the four lines $A'P, B'P, C'P, D'P$ which form a pencil through P. The cross-ratio of this pencil can be shown not to depend upon the choice of P (the angles at P do not depend upon P by Euclid III, 21 – angles in a segment are equal) and so we can take this to define the cross-ratio of the four points on the circle. So finally by taking P as the centre of inversion it is clear that the cross-ratio of the four points on the circle is the same as that of those on the line. (Incidentally, the cross-ratio on any other conic such as an ellipse is defined in the same way.)

We have now introduced the key properties of inversion which was invented by Jakob Steiner whose porism we will return to in a moment.

Coaxial circles

CONSIDER TWO CIRCLES through two points C and D as shown below. Take any point P on CD (not between C and D) and consider the tangents to the two circles meeting them at T and U. Now we have

$PT^2 = PC \times PD = PU^2$ using Euclid III, 36 twice

So the tangents PT and PU have the same length. So the circle centre P and radius PT will be orthogonal to both circles and indeed to all circles through C and D.

But we can do this with any point P and so obtain a whole series of circles whose centres lie on the line CD each of which is orthogonal to all the circles through both C and D.

The result is the very pretty arrangement shown opposite above. There are two sets of circles. The circles of one set meet each other in C and D (we shall call this set the M-set – or meeting set). The other set (the N-set – or not meeting set) do not meet each other and have their centres on the line CD. All the circles of one set are orthogonal to all circles of the other set.

The two straight lines in the diagram from which the tangents to all circles of one set are equal are known as the radical axes. The two sets are known as coaxial sets.

Now consider what happens if we invert the whole diagram with respect to a circle whose centre is D. For convenience we can take the circle of inversion as the circle with centre D that passes through C so that C inverts into itself.

We recall that a circle through the centre of inversion becomes a straight line not through the centre of inversion. So the M-circles through D become straight lines and moreover they all go through C since C remains unchanged. The N-circles on the other hand invert into another set of circles. However, since angles are preserved, these circles have to be orthogonal to the set of straight lines though C. This means of course that they are concentric and all have centre C. The result is shown below opposite.

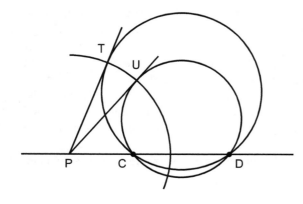

The tangents PT and PU from any point P to any circles through C and D have the same length.

So a circle centre P with radius PT is orthogonal to all circles through C and D.

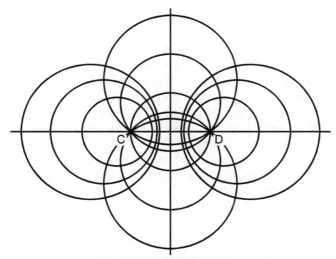

Two sets of coaxial circles.

As a consequence, if we take *any* two circles that do not intersect (any two N-circles) and invert them about the point D then the result is a pair of concentric circles. Of course, we have to find the point D but that is easy. Just draw a circle orthogonal to the two given circles and then that meets the line joining the two centres in C and D. We can also invert about C; again the inverse circles will be concentric.

And now at last we can easily prove Steiner's porism.

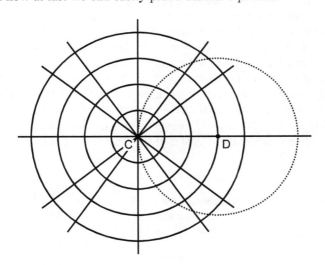

Inverting the two sets of circles with respect to D results in a set of concentric circles about C and a set of lines through C.

Proof of Steiner's porism

SUPPOSE we have two circles one inside the other so that they do not intersect and that we can draw a chain of circles touching them and each other so that the chain neatly joins up. Then Steiner's porism states that the chain always joins up no matter where we start the chain. And contrariwise, if the chain doesn't join up then it never will no matter where the chain is started.

We have just seen that any two circles can be inverted into a pair of concentric circles. If we do this then the chain of circles will transform into a ring of equal circles between the two concentric circles as shown below. Now in the transformed picture the ring will obviously work wherever it is started because being all the same size the little circles can be rotated like ball bearings. The corresponding circles in the inverted figure on the left will then expand and contract as they rotate between the two fixed circles. So that proves it. Amazing.

(For convenience I will refer to the equal circles between the concentric fixed circles as a ring of circles and to the variable inverted ones as a chain of circles.)

The centre of inversion is inside the circles as marked below. Note that since the points of contact of the ring of circles on the right are themselves on a circle, then the points of contact of the transformed chain on the left are also on a circle. But the same doesn't apply to the centres since, as noted earlier, the centre of a circle does not invert into the centre of the inverted circle. In fact the centres of the chain lie on an ellipse – the proof is left to the reader.

It is amusing to consider what happens if the centre of inversion is in different places. And then to consider how the system behaves as the circles rotate. For reasons which will become apparent when we deal with an analogous system in three dimensions, we will deal with the case of six circles although similar behaviour occurs with any number. Note that with six, the circles in the original ring have the same radius as the inner fixed circle – and consequently the outer fixed circle has three times the radius as the inner one. The simple case with the point of inversion inside the inner circle is shown opposite with the chain of transformed circles in four positions. The first corresponds to the position of the original ring of equal circles; the others correspond to rotation of the original ring by 15°, 30° and 45° respectively. In this example, taking the inner circle as having radius 1 and centre at the origin, the point of inversion is

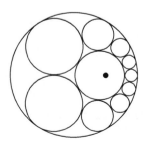

The figure on the left can be inverted into that on the right.

The blob is the centre of inversion.

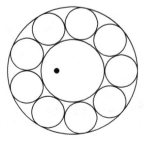

7 Harmony of the Spheres   167

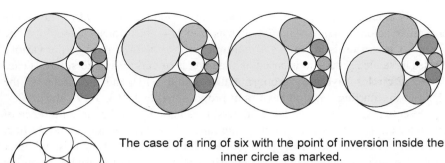

The case of a ring of six with the point of inversion inside the inner circle as marked.

Four positions of the inverted system are shown corresponding to the original ring being rotated in steps of 15°.

the point (−0.5, 0.0). Note that the inner circle of the original becomes the outer circle of the inverted system and vice versa.

As the point of inversion approaches the circumference of the inner circle, the inverted inner circle becomes very large and the circles of the chain expand and contract a great deal as they tour around. Two positions are shown below in a fairly extreme situation – these correspond to the first two positions in the previous example, that is at the starting position and with 15° rotation.

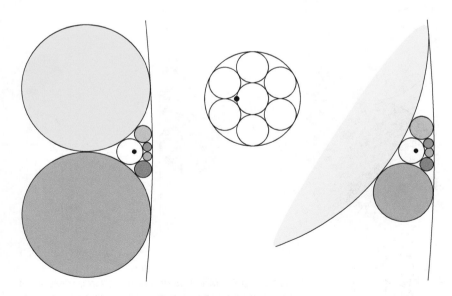

When the point of inversion is very close to the inner circle, the inverted circle becomes huge. Two positions of the chain are shown corresponding to the original ring and it rotated by 15°.

168   Gems of Geometry

This example has the point of inversion at about (–0.99, 0.0). The figure chosen is actually 4/7 times √3 which is 0.98974.... Curiously enough, if the centre of inversion is at fractional multiples of √3 then the radii of the various circles have nice proportions based on the inverse of various integers. In this case the fixed circles have radii proportional to 1 and 1/131 and the circles of the chain in the starting position have radii proportional to 1/363, 1/195 and 1/27.

Eventually, when the centre of inversion is actually on the circumference of the inner circle, that is at (–1.0, 0.0), the inner circle inverts into a straight line. The chain of circles still manage to tour around the inverse of the original outer circle. However, each becomes momentarily a straight line (essentially becoming a circle of infinite radius) when it is exactly opposite the line which is the image of the original inner circle. This happens when the corresponding circle of the original ring just touches the centre of inversion as it rotates around.

Three positions are shown below corresponding to the starting position and with rotations of 15° and 30° respectively.

Next consider what happens when the point of inversion moves a bit further so that it is in the gap between the two fixed circles. The inner circle now inverts into a circle which does not surround the image of the outer circle. In other words

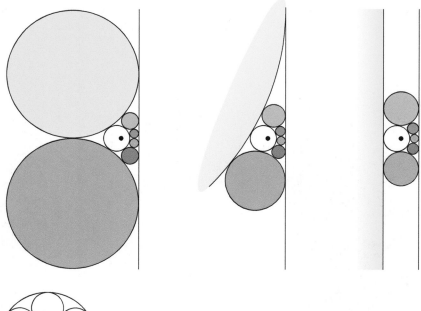

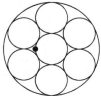

When the point of inversion is on the circumference of the inner circle, the inverted inner circle becomes a straight line. Three positions of the inverted system are shown corresponding to the original ring being rotated in steps of 15°.

7 Harmony of the Spheres

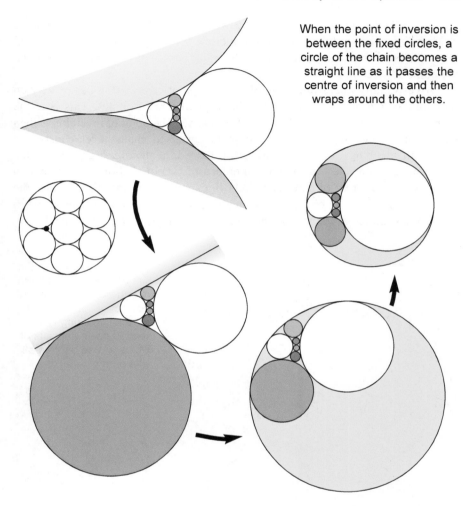

When the point of inversion is between the fixed circles, a circle of the chain becomes a straight line as it passes the centre of inversion and then wraps around the others.

the two fixed circles no longer define an annular space. At the starting position the centre of inversion is not within any of the circles of the ring and the inverted chain consists of six circles touching each other in the usual way.

The diagram shows the case when the point of inversion is at about (−1.299, 0.0). The exact value is in fact 3/4 times √3. Again the radii have nice ratios. The fixed circles have radii proportional to 1/11 and 1/39 and the circles of the chain in the starting position have radii proportional to 1/147, 1/75 and 1/3.

Now, as the ring rotates, the behaviour is perhaps rather surprising. As each circle passes the point of inversion, its image becomes a straight line and then as it progresses further its image becomes a circle wrapping the others. The diagram above shows four positions: the usual starting position, and then rotated such that the first circle becomes a straight line (at about 4.5° in this example) and then at

15° and 30°. As rotation continues the wrapping circle eventually becomes a straight line again and then changes to a circle touching the others in the usual way. The process repeats of course as the other circles of the ring pass through the point of inversion.

An especially interesting case arises if the point of inversion lies exactly on the circle of the points of contact of the original ring of circles. This is at the point (−1.732..., 0.0) where 1.732... is of course √3. The point of inversion is then never outside the ring of circles and so the inverted chain never consists of circles all touching each other externally. Indeed, at the starting position, the centre of inversion lies on the circumference of two circles of the ring and so they invert into a pair of parallel straight lines. As the original ring rotates, the two lines become huge circles one wrapped around all the others. The diagram shows the situation at the starting position and at angles of 7.5°, 15° and 30°.

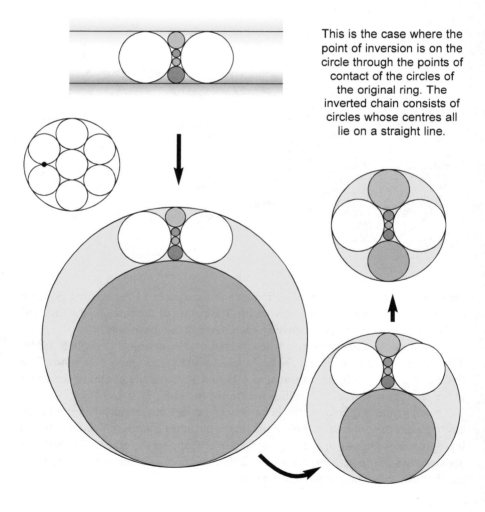

This is the case where the point of inversion is on the circle through the points of contact of the circles of the original ring. The inverted chain consists of circles whose centres all lie on a straight line.

Note that the fixed circles are now equal and the centres of the inverted circles lie on the straight line midway between the fixed circles. Let us follow the green circle which in the diagram opposite starts as the bottom straight line which corresponds to its centre being at infinity at the bottom of the page. As the system progresses, the green line turns into a huge circle which slowly contracts and as it does so its centre moves up until the circle is at its smallest position and its centre is between the fixed circles. The centre continues to move up and the circle grows until it once more becomes a straight line only this time it is the upper straight line. The centre now flips from infinity at the top to infinity at the bottom and now the green circle becomes the huge one that wraps around all the others. Its centre now moves up from infinity at the bottom and the circle shrinks until its centre is once more between the two fixed circles only this time it surrounds all the circles. Finally, the centre progresses upwards and the green circle grows until it once more becomes the bottom straight line which is where we started.

The various systems can be classified according to how the centres of the chain of inverted circles behave. When the point of inversion is inside (or outside) both fixed circles then the centres of the chain lie on an ellipse. When the point of inversion is on the circumference of a fixed circle then the centres of the chain lie on a parabola. When the centre of inversion is between the fixed circles then the centres of the chain lie on a hyperbola. Finally, when the centre of inversion is on the circle of contact, then the centres of the chain lie on a straight line as we have just seen. The diagram below illustrates this categorization.

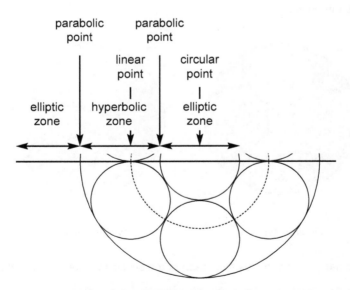

The centres of the circles of the inverted chain lie on a conic whose type depends upon the position of the centre of inversion.

172   Gems of Geometry

Note that if the centre of inversion moves further out then we get the same kinds of systems again but in the reverse order. Indeed each inverted system corresponds (apart from an overall scale factor) to inverting the original ring at two different places. Thus inverting at (−1, 0) which is on the circumference of the inner fixed circle gives rise to the same kind of system as inverting at (−3, 0) which is on the circumference of the outer fixed circle. Similarly, inverting at (−0.5, 0) which is inside both fixed circles gives the same system as inverting at (−6, 0) which is outside both circles. The only difference is that in the inverted system the inner and outer circles change places. The reader might recall when discussing the inversion of coaxial systems that both points C and D would invert the coaxial system into a concentric system and these correspond to the twin points here.

Everything we have discussed applies in general no matter how many circles are in the ring. We chose six because that configuration will turn up again in three dimensions in a moment.

An interesting special case is when there are just four circles in the ring. This results in a very symmetric configuration in which each circle including the two fixed circles touches four of the other five. If we invert this system then naturally we get another system with six circles in which each touches four of the others. And in fact if we invert at the point on the circumference of the circle through the points of contact of the circles of the ring (the linear point in the categorization diagram), then the inverted system is identical except that two pairs of circles have interchanged their places.

Well that really concludes the story of Steiner's porism. And now to move into three dimensions.

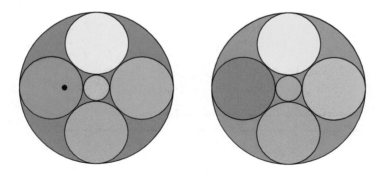

Original system.   Inverted system.

If the ring has just four circles then each of the six circles meets four of the others.

The inverted system with respect to the point shown on the circle of contact points is exactly the same with two pairs of circles interchanged.

Soddy's hexlet

THERE IS an amazing analogue of Steiner's chain in three dimensions which was first discovered by Sir Frederick Soddy (1877–1956). Soddy is mainly famous for his work on isotopes (he invented the word isotope) and he was awarded the Nobel prize in 1921. As well as being a brilliant chemist, he was also an enthusiastic geometer and he discovered the hexlet in 1932.

Consider three spheres touching each other. There is a gap between them and the question arises as to whether we can form a closed chain of spheres passing through this gap such that each sphere touches the three fixed spheres and the adjacent spheres in the chain. By analogy with Steiner's porism we might imagine the answer to be: "Yes, provided the spheres have certain positions." But the amazing fact is that no matter how the three spheres are placed, it is always possible to have a closed chain and moreover there are always exactly six spheres in the chain. Hence the term hexlet.

The proof that this is the case is remarkably simple. We use inversion once more but this time in three dimensions. The rules for inversion in three dimensions are as one would expect. Spheres invert into spheres except that a sphere through the centre of inversion inverts into a plane (just as in two dimensions a circle through the centre of inversion inverts into a line).

Now invert the three fixed spheres with the point of contact between two of them as the point of inversion. The two spheres through the point of inversion (A and B say) become parallel planes (A' and B') and the third sphere (C) becomes a sphere (C') between them and touching them. Clearly we can now place a ring of six equal spheres around the fixed sphere and these will all touch the two planes as shown below. (Just imagine seven billiard balls on a table with six around a central one and then place a sheet of glass on top.) The points of contact of the six spheres of the chain with the planes form the vertices of a hexagon. If we now invert the system back then we get a chain of six spheres touching the three fixed spheres with which we started. QED

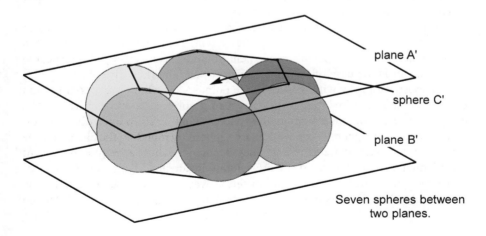

Seven spheres between two planes.

174   Gems of Geometry

(For convenience I will again refer to the circle of six spheres between the planes as a ring of spheres and the inverted spheres as a chain of spheres.)

We can classify the different configurations possible by considering various points of inversion of the system of two planes and seven equal spheres just as we did with Steiner's circles. Before doing so, it is interesting to note that a further sphere (D', not shown) surrounds the six spheres of the ring. However, this sphere intersects the two planes and so in any inverted system the sphere D will intersect the spheres A and B.

We will start by supposing that the point of inversion lies centrally between the two planes so that it is in fact in the plane passing through the centres of the seven spheres. One consequence of this is that the spheres A and B are of the same size. Another consequence is that by considering the central plane we see that the cross-section of the spheres reveals the Steiner configuration we have just discussed in detail. The inner sphere (C') corresponds to the inner circle and the outer sphere (D') corresponds to the outer circle and of course the ring of spheres corresponds to the ring of circles.

As a first example, suppose that the centre of inversion is outside the original ring of spheres. This corresponds to the case where the centre of inversion is outside the system of circles and so as the ring of spheres rotate, they do not hit the centre of inversion and so nothing unusual happens to the corresponding chain of spheres as it rotates.

Three views of such a system are shown opposite above. We assume that the fixed spheres are transparent and that the spheres of the chain are coloured and solid. The first view is in the direction of the line joining the centres of A and B so that A and B are superimposed in the projection. The other two views are at right angles. The sphere D is shown in the first view only.

In this example the centre of inversion is at the point $2\sqrt{3}$. The sizes of the spheres A, B, C and D are proportional to 1/2, 1/2, 1/11 and 1/1 respectively and the spheres of the chain are proportional to 1/3, 1/15 and 1/27.

Note an interesting corollary. If we start with any Steiner chain of six circles and consider the corresponding spheres, then we can always place a pair of equal spheres A, B to touch the spheres of the chain and each other.

In the case of the Steiner circles we saw that two complementary points of inversion essentially give rise to the same system. Thus inverting with respect to the point $(x, 0)$ gives the same arrangement of circles as inverting with respect to $(-3/x, 0)$. But in the case of the spheres there is an important difference.

Suppose we invert with respect to the point $-\sqrt{3}/2$ which is the complement of $2\sqrt{3}$. We are now inverting with respect to a point inside the sphere C'. The chain of spheres remains the same, the spheres C and D swap over which means that the important sphere C is now the large sphere surrounding the chain, but the spheres A and B are quite different. They become much smaller (one-quarter of their former size to be precise) and are actually inside the large sphere C. So here is a major difference reflecting the different ways in which three spheres can touch each other. The result is shown opposite below. The spheres A and B touch each other through the central gap in the chain.

7 Harmony of the Spheres 175

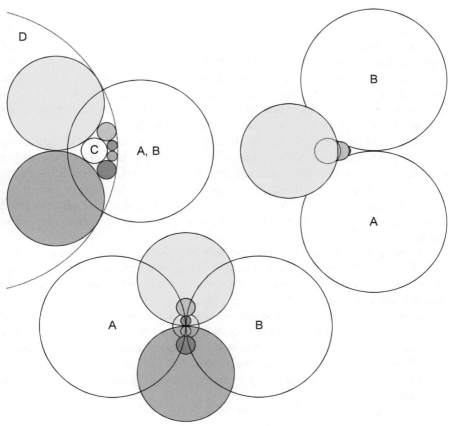

(*above*) Three orthogonal views of six spheres forming a chain around the three fixed spheres.

(*below*) The same six spheres forming a chain around the three fixed spheres. But the fixed spheres A and B are inside the fixed sphere C.

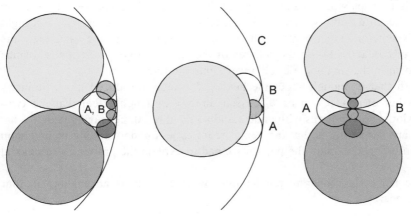

Perhaps it would be easier to understand what is going on if we consider the very simple case of inversion with respect to the origin. In Steiner's porism this is pretty boring because the configuration simply inverts into itself (the inner and outer circles change places of course). But in the hexlet, the two planes invert into equal spheres A and B and the inner sphere C' becomes a sphere C encircling them symmetrically. The ring inverts into a chain of equal spheres which run around the equatorial gap rather like a curious ball bearing. The top diagram opposite shows the usual three views of this configuration.

Just as in Steiner's porism we obtain different classes of configurations according to the location of the centre of inversion. If it is between the planes but outside the inner sphere then the fixed spheres touch externally. If it is inside the inner sphere then the two fixed spheres corresponding to the plane are inside the third sphere as above. If it lies outside the planes then again we get two fixed spheres inside the third but this time the outer sphere corresponds to one of the planes.

If the centre of inversion lies between the planes in such a way that the spheres of the ring pass through it as they rotate, then the spheres of the chain become planes and then wrap around the other spheres in turn in much the same way that the circles behaved in Steiner's porism.

In particular, if the centre of inversion is exactly on the circle of points of contact between the spheres of the ring then the three fixed spheres are the same size and the spheres of the chain then move so that their centres are on a straight line. They take it in turn to become planes and then wrap around the entire system. Three views of this pretty symmetrical configuration are also shown opposite.

Other variations are much as in Steiner's porism. If the point of inversion lies on the surface of the inner fixed sphere then of course that sphere inverts into a plane and the fixed spheres become two spheres touching that plane. If the point of inversion is not on the central plane then the fixed spheres A and B are not the same size but this does not much change the general features of the resulting configuration.

It is perhaps worth mentioning that the centres of the spheres of the chain always lie on a conic which can be an ellipse, parabola, hyperbola or straight line. Except for the case where the three fixed spheres are the same size, the points of contact between the spheres of the chain always lie on a circle since they do so in the ring and a circle inverts into a circle.

Well that is enough about the hexlet. We have introduced it via Steiner's porism but clearly it can be discussed directly although I feel that the two-dimensional porism helps with understanding the three-dimensional hexlet. What is perhaps surprising is that the hexlet requires less understanding of inversion since in order to explain the porism we had to explain the properties of coaxial circles.

We leave the reader to ponder over whether there is an analogy in four dimensions.

7 Harmony of the Spheres 177

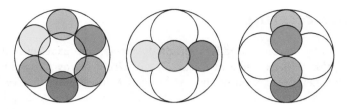

Three views of the case where the centre of inversion is at the origin. The two equal fixed spheres are centrally placed in the third fixed sphere.

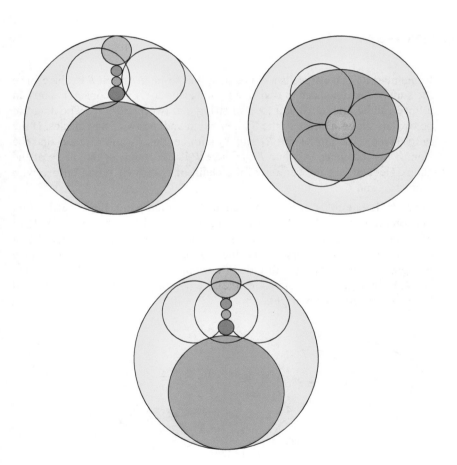

Three views of the case where the three fixed spheres are all the same size.

The yellow sphere surrounds them all and has been made transparent otherwise we couldn't see much! The three fixed spheres are transparent as usual.

178 Gems of Geometry

Further reading

APART FROM READING Soddy's original papers, an excellent alternative description will be found in *Excursions in Geometry* by C Stanley Ogilvy. He gives a fuller description of the background to inversion than we have but says rather less about Steiner's porism and goes straight for a description of the hexlet. The book also covers aspects of projective geometry, the golden number and other pretty topics. A lovely little book which encouraged me to explore the beauties of the hexlet.

Exercises

1 Remember that an ellipse is a curve such that from any point P of the ellipse the sum of the distances to two fixed points (the foci) is a constant. Thus an ellipse can be drawn using a loop of string and two fixed pins. The diagram below left shows a circle of the Steiner chain and the two fixed circles. If the radii of the fixed circles are R_1 and R_2 with centres C_1 and C_2 then find the sum of the distances PC_1 and PC_2 where P is the centre of the circle of the chain. Deduce that the centres of the chain of circles lie on an ellipse with foci at C_1 and C_2.

2 Suppose three circles A, B, C intersect at a common point. Suppose also that the common chord of A and B is a diameter of C and that the common chord of A and C is a diameter of B. Prove that the common chord of B and C is a diameter of A. Use inversion and the theorem that the three altitudes of a triangle are concurrent (an altitude of a triangle is a line through one vertex and at right angles to the opposite side). Remember that angles are preserved by inversion. See below right.

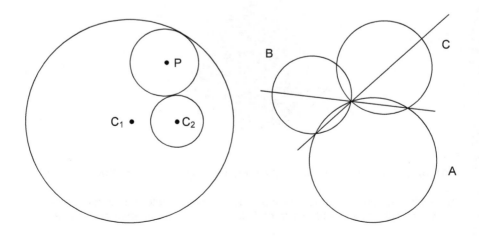

8 Chaos and Fractals

THE TWENTIETH CENTURY saw an upheaval in our understanding of the mechanics of the universe and the foundations of mathematics. This lecture looks at two aspects of strange modern mathematics with a clear geometrical interpretation and the subsequent lecture looks at aspects of our understanding of the physical universe as modified by Einstein's theories of relativity.

Shaken foundations

AT THE END of the nineteenth century, it seemed clear that our understanding of the universe was more or less complete at least with regard to the laws of physics. It was a world of order and predictability. Newton's laws prescribed the motion of matter in an accurate manner. Maxwell had devised his equations describing electricity and magnetism. And Whitehead and others were working on the foundations of mathematics and seemed to have placed it on a solid footing.

There were a few loose ends. The orbit of the planet Mercury didn't seem quite right; black body radiation was a bit odd and in mathematics there were things like Fermat's last theorem still to be proved. But these were surely just loose ends that needed tidying up. Generally speaking, it was assumed that it was a clockwork universe. Given the data describing the position and momentum of every particle then we could, in principle if not in practice, compute the future. This raised awkward philosophical questions of freewill but then philosophers always ask awkward questions anyway.

The twentieth century destroyed all this. First of all, Einstein introduced his theories of relativity which showed that straight lines weren't even straight and Newton's laws weren't quite right. And then Dirac, Schrödinger and others introduced quantum mechanics which threw predictability away so that it was not a clockwork universe after all. And to make matters much worse, Kurt Gödel shook the very foundations of pure mathematics itself by showing that some things could neither be proved nor disproved.

Much new curious mathematics emerged in the second half of the twentieth century (although the foundations were laid about a century earlier) which in a sense is about coping with complexity which defies a traditional approach.

Two aspects which we will now briefly explore are the ideas of fractals which have the odd property of being characterized by a dimension which is not a whole number and then so-called chaotic systems which arise from non-linear dynamical equations which cannot be solved using the approaches of the Victorian mathematicians.

Fractals

IF WE LOOK at a coastline on a map then it is clear that it is not smooth. There are bays and estuaries and promontories and creeks and so on. If we look at a large scale map then further irregularities appear, big bays often have small bays in them, promontories have little promontories, creeks have further tiny creeks off them. And of course if we walk the coastline then we discover even more fine detail. Structures which exhibit a similar pattern whatever the scale are said to be self-similar.

It is impossible to define the length of a coastline without giving details of how we are going to measure it. If we use a small scale map (say 1:100,000) and mark off the coastline in 1 km lengths with a pair of dividers then we might find a length of 100 km. If we measure the same coastline on a large scale map (say 1:10,000) and mark it off in 100 metre lengths then it will be longer because we will go in and out of bays which were ignored on the first map. It might be 120 km. If we walk the coastline and measure it with a 10 metre chain then it will be longer still. And if we crawl along it with a metre rule then it will be even longer.

A formal mathematical example of this idea is given by the curve devised in 1904 by Helge von Koch (1870–1924), a Swedish mathematician. In the first stage below we have a line consisting of four straight sections each of unit length but with the middle two sections around the outline of an equilateral triangle. The length of the line is four units but the distance between the endpoints is three units. We now replace each of the four sections by a copy of the original reduced by a factor of three and this gives stage 2.

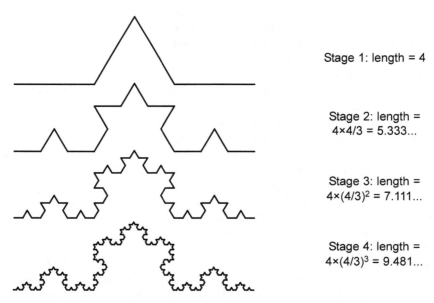

Stage 1: length = 4

Stage 2: length = 4×4/3 = 5.333...

Stage 3: length = 4×(4/3)² = 7.111...

Stage 4: length = 4×(4/3)³ = 9.481...

Four stages in the development of the Koch curve.

We then repeat the process for ever. It is clear that at each stage the length of the line is increased by a factor of 4/3. This continues indefinitely and so the ultimate line (if it could be drawn) would be infinitely long. The line at stage 100 is over 7 billion units long (and good old-fashioned billions of 10^{12} at that).

Nevertheless, although the line becomes infinitely long it is still contained very much in the original area of the early stages – it just gets fuzzy. Moreover, it really is a line still since if we cut it by a point then it becomes two parts.

Fractional dimensions

WE ARE FAMILIAR with the idea of one, two, and three dimensions. A key characteristic of figures with these dimensions is that if we double a linear dimension then the size of the overall object is increased by 2, 4, or 8 respectively, that is 2 raised to the power given by the dimension. Thus, if we have a square of side 1, then a square of side 2 can be thought of as composed of 4 squares of side 1. And similarly a cube of side 2 can be composed of 8 cubes of side 1.

We can also look at this in reverse. If a two-dimensional object is subdivided into N smaller ones then the size of the original will be $\sqrt{N}$ times that of the smaller ones (of course N had better be a square number such as 25 in order to make this practicable). And if a three-dimensional object is subdivided into N smaller ones then the size of the original will be $\sqrt[3]{N}$ times that of the smaller ones. So the ratio r between the original and the subdivided version is related to the dimension and the number of subdivisions by

$r^D = N$ the dimensions power of the ratio is the number of subdivisions

If we take logarithms we obtain an expression for the number of dimensions D in terms of the number of subdivisions N and the ratio r, thus

$D = \log N / \log r$ the dimension defined in terms of the scaling

Let us just check this for consistency. If we have a cube of side 10, then we can divide it into 1000 cubes of side 1. So N is 1000 and r is 10. For convenience we

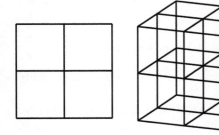

A square can be subdivided into 4 squares of half the linear size.

A cube can be subdivided into 8 cubes of half the linear size.

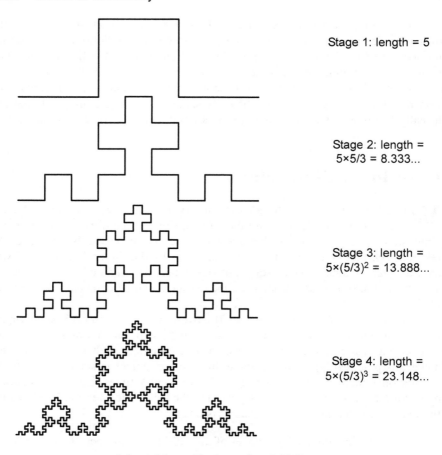

A fractal line with dimension 1.4649....

can use logarithms to base 10 (the base doesn't matter) and find that $\log N = \log 1000$ is 3 and $\log r = \log 10$ is 1 giving $D = 3/1 = 3$ as expected.

We can now define the dimensionality of a fractal using this very same formula.

In the case of the Koch curve, we divided it into four parts at each stage and each part was 1/3 of the linear size of the original. So N is 4 and r is 1/3. We obtain

$D = \log 4 / \log 3 = 1.2618...$ dimensionality of Koch curve

It has been found by measurement that a natural coastline has a dimensionality of about 1.2.

We can apply the same subdivision process that we did with the Koch curve to other fragmented lines. The diagram above shows the effect of using a square

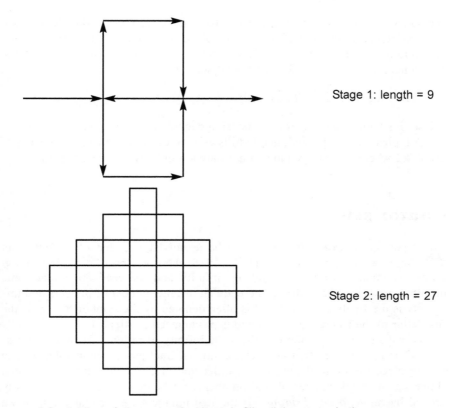

A fractal line of dimension 2 ultimately fills all the space in the square.

outline rather than the triangular outline in the Koch curve. This divides the line into 5 parts and so we obtain

$$D = \log 5 / \log 3 = 1.4649...$$

If we now make the square subdivision on both sides so that the line goes around one square and then around the other we obtain a division into 9 segments. The dimension calculation then becomes

$$D = \log 9 / \log 3 = 2$$

Rather surprisingly the dimension is now exactly 2 and indeed the ultimate line visits all space within a square whose opposite vertices are the original ends of the line. The line has now in fact become a square and this is why the dimension is 2.

The same ideas apply to surfaces. Thus we can take a triangle, decompose it into four triangles and then replace the centre one by a tetrahedron. The original triangle had area 4 and the new figure is composed of six equilateral triangles and so has area 6. Using this figure as a baseline, clearly each triangle of it can now be replaced by a half size copy of itself and this can be continued indefinitely – note that we say half size since it is the linear ratio that matters. The dimension calculation for this fractal surface is

$$D = \log 6 / \log 2 = 2.58449...$$

and so this bumpy surface has a dimension considerably more than 2.

A typical landscape with gentle hills and valleys has a fractal dimension of about 2.2 whereas a craggy mountainous area might have a dimension of 2.5.

Cantor sets

AN IMPORTANT CONCEPT which will be useful later is that of a Cantor set named after George Cantor (1845–1918) who was born at St Petersburg. Cantor is famous for his research on infinity and the realization that some infinities are larger than others. There are an infinite number of integers and this infinity is the smallest infinity and is often denoted by the symbol $\aleph_0$. ($\aleph$ is the first letter of the Hebrew alphabet and is pronounced as Aleph.)

Perhaps surprisingly there is the same infinity of rational fractions such as 1/2, 2/3 and so on; we often say that the rational fractions are countable. We can in fact place the rational numbers in order and match them with the integers. However, the total number of real numbers is a bigger infinity. This is easily proved by *reductio ad absurdum*. If the real numbers were countable then we could put them in order and match them to the integers. Let us suppose that we have done this for the numbers between 0 and 1 and write out their decimal expansion thus

1    $0.a_1 a_2 a_3 a_4 a_5...$
2    $0.b_1 b_2 b_3 b_4 b_5...$
3    $0.c_1 c_2 c_3 c_4 c_5...$
4    $0.d_1 d_2 d_3 d_4 d_5...$

We now fabricate a number which is different from the first number in the first place and is different from the second number in the second place and so on (this is a technique called diagonal slicing). This number is clearly not in the list because it differs from them all somewhere. And so the list wasn't complete and so the numbers could not be matched with the integers and therefore there are more real numbers than integers. Well, I have got off the subject, but I just wanted to mention some of the clever ideas and techniques introduced by Cantor.

A Cantor Middle Thirds Set is obtained by taking all the real numbers from zero to one and then removing the middle third (but not removing its end points). The result is two sections of numbers from 0 to 1/3 and from 2/3 to 1. We then do the same thing with these two sections and repeat the process for ever.

Creating the Cantor Middle Thirds Set or Cantor Dust.

It is clear that the sum of the lengths of the lines remaining is multiplied by 2/3 at each stage and so goes to zero. However, although the sum of the lengths is zero, there are still some numbers left and in fact there are an infinite number of them left. Moreover, each one is quite isolated and for this reason the set is sometimes called Cantor Dust. It is another example of a self-similar structure.

Note that it is very important that we do not remove the end points at each stage. In fact these end points that are left form the dust. So the set contains the numbers 0, 1, 1/3, 2/3, 1/9, 2/9, 7/9, 8/9, 1/27, 2/27, 7/27, 8/27, 19/27, 20/27, 25/27, 26/27 and so on.

Population growth

IN LECTURE 1 we discussed Fibonacci's rabbits and noted that their population grew each month by a factor which converged upon the golden number τ. So in due course the population in month $N+1$ which we can call P_{N+1} is given in terms of the population in month N by

$P_{N+1} = \tau \times P_N$ exponential rabbit growth

We can alternatively write this as

$f(x) = \tau \times x$ f gives the population in terms of its previous value

In other words $f(x)$ is a simple function which takes the population in one month and gives the population in the next month.

With this model the population grows without bounds as shown in the graph below. The world eventually becomes solid rabbit.

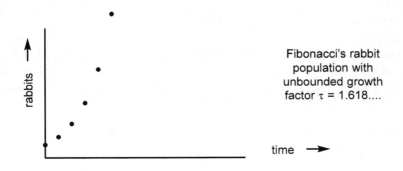

Fibonacci's rabbit population with unbounded growth factor τ = 1.618....

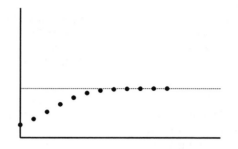

Population with growth factor $\tau = 1.618$ stabilizes at $\tau^{-2} = 0.382....$

In practice of course the growth of the rabbit population is limited because of running out of food (grass, carrots, etc). We can express this by modifying the function $f(x)$ in some way so that it cannot get excessively large. An obvious way to do this is to multiply it by some factor which gets smaller as x gets larger. So we might consider

$f(x) = \tau \times x \times (1-x/N)$ growth modified with limiting constant N

What happens now is that as the population approaches N, the factor $(1-x/N)$ restricts the growth and eventually halts it. (Note that we are now dealing with fractions of pairs of rabbits! We can overcome this by assuming that the population is very large and that we are dealing with percentages.) Suppose N is 100 and the value of x at the start is 10, then the sequence of values in successive months (to two decimal places) is

10.00, 14.56, 20.13, 26.01, 31.14, 34.70, 36.67, 37.57, 37.95, 38.10, ...

The population stabilizes at 38.1966.... We can easily compute this because $f(x)$ then equals x so

$x = \tau \times x \times (1-x/100)$ condition for static population, which gives
$x = 100 \times (1-1/\tau)$ and then using $\tau^{-1} + \tau^{-2} = 1$, we get
$x = 100 \times \tau^{-2} = 38.1966...$

Well I suppose we would naturally expect the population to stabilize in this way. The graph above shows the sequence and the steady limit.

Biologists have known for a very long time that populations are often not stable but can jump about in alarming ways. Sometimes this is attributed to freak conditions such as a drought but it is in fact inherent in the simple model we have just been using. For convenience it is best to get rid of the number N and to work entirely in terms of fractions. The general equation then becomes

$f(x) = \mu \times x \times (1-x)$ which is usually written as
$f(x) = \mu x(1-x)$

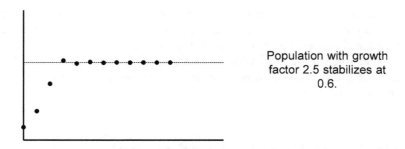

Population with growth factor 2.5 stabilizes at 0.6.

For some reason it is traditional to use the Greek letter μ (*mu*) for the growth factor in this equation. We have just been dealing with the case where the growth factor μ = τ = 1.618... and we have seen that this results in a stable population. Let's try some other values of the growth factor starting at 0.1 each time (which corresponds to starting at 10% as before).

Values of μ less than 1 (or indeed equal to 1) result in the population dying out and so are rather uninteresting. Values between 1 and 2 behave much like τ and the population steadily approaches a limit which is determined by setting *f*(*x*) equal to *x* so that we have

$x = f(x) = \mu x(1-x)$ which can be rearranged to give

$x = 1 - 1/\mu$ fixed point for growth factor μ

Values between 2 and 3 behave somewhat differently – although the population does stabilize at a level given by the same formula it always overshoots first and then oscillates about the limit as it settles down. The case of μ = 2.5 is shown above.

However, as μ approaches 3 it takes a long time to settle down. The case where μ = 2.9 is shown below. And when μ is equal to 3 it seems as if it will never settle down but it does eventually at 0.666....

The limit point in the cases we have just been discussing is known as a fixed point because if we start from it then the next value is the same. In these cases it is also known as a stable point (or an attracting point) because if we disturb the system slightly then it eventually goes back to the same fixed point. Note that

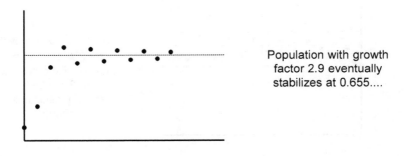

Population with growth factor 2.9 eventually stabilizes at 0.655....

zero is also a fixed point since if we start at zero then we stay at zero since $f(0)$ is always 0 no matter what the value of μ. Moreover, if μ is less than or equal to 1, then the fixed point at zero is stable but for the case of μ greater than 1 it is not stable since if we start with a value slightly different from zero then the population grows away from zero. In these cases we say that zero is an unstable point (or a repelling point).

Double, double, boil and trouble

AND NOW to return to other values of μ. When μ becomes a bit more than 3, the behaviour becomes rather different. It doesn't actually settle down at all but ends up by oscillating between two values. We say that it has a stable period of two. The period is stable because if we disturb it slightly then it settles down again to the same period of two values. Note that the original fixed point at $1-1/\mu$ is still there but it is no longer stable and so if we start just slightly away from it then the population moves into the stable cycle of two.

The case of $\mu = 3^{1}/_{6}$ is shown below. If we start at 0.1 as before then it eventually settles down into a cycle between the two points $10/19 = 0.526...$ and $15/19 = 0.789...$. This is shown in the upper graph. Moreover, the unstable fixed point is at $13/19 = 0.684...$. The lower graph illustrates what happens if we start near this fixed point at the point 0.7. It oscillates around the fixed point but gradually diverges until it eventually settles into the two cycle as before.

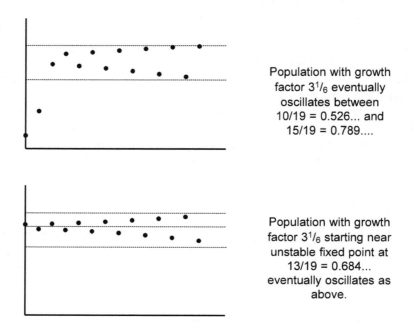

Population with growth factor $3^{1}/_{6}$ eventually oscillates between $10/19 = 0.526...$ and $15/19 = 0.789....$

Population with growth factor $3^{1}/_{6}$ starting near unstable fixed point at $13/19 = 0.684...$ eventually oscillates as above.

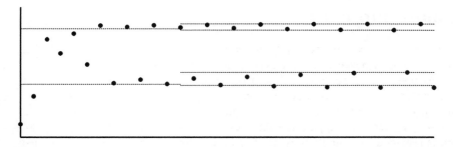

Population with growth factor 3.5 appears to be going to oscillate between 6/7 = 0.857... and 3/7 = 0.428... but finally settles into a four cycle.

The equation for determining the points of the two cycle is obtained by applying the function f twice so we have

$f(f(x)) = x$ which eventually gives the indigestible

$\mu^3 x^3 - 2\mu^3 x^2 + \mu^2 x + \mu^3 x - \mu^2 + 1 = 0$

This is not quite so awful as it seems because it includes the factor $(\mu x - \mu + 1)$ corresponding to the fixed point and taking this out we get the quadratic equation

$\mu^2 x^2 - \mu(\mu+1)x + (\mu+1) = 0$ equation for two cycle

and the two roots of this equation give the two values for the two cycle.

If we increase μ a bit more then another change occurs. The stable two cycle itself becomes unstable and a new stable four cycle is introduced. As happened with the stable fixed point, the two cycle is still there but simply becomes unstable. This change occurs when $\mu = 1 + \sqrt{6} = 3.449....$ See Appendix F for more details of the values of the two cycle.

The case of $\mu = 3.5$ is illustrated above. The unstable fixed point is at exactly 5/7 = 0.714... and the unstable two cycle is between 6/7 = 0.857... and 3/7 = 0.428.... The new stable four cycle goes around the four points 0.875, 0.826..., 0.382..., 0.500. Curiously enough, starting from 0.1, it first oscillates around the unstable two cycle before settling into the stable four cycle.

As we increase μ more and more the cycles double again and again with ever increasing frequency. At each stage a new stable cycle of period 8, 16, 32, ... is introduced and the previous stable cycle becomes unstable. Thus at $\mu = 3.55$, there is an 8-cycle; at $\mu = 3.566$, there is a 16-cycle; at $\mu = 3.569$, there is a 32-cycle; and at $\mu = 3.5696$, there is a 64-cycle.

A dramatic change occurs at around $\mu = 3.57$. The sequence of values reveals little discernable structure; the values jump about all over the place and never repeat. We say that the behaviour is chaotic. Some areas of values seem to be omitted but there are no stable cycles. Of course the two, four, etc. cycles still exist but they are all unstable.

Chaos and peace

THE CHAOTIC PATTERN generally holds up to $\mu = 4$ but strangely enough there are areas of peace where the system again reveals stable behaviour.

At $\mu = 3.82$, chaos reigns, but at $\mu = 3.83$ the behaviour is quite different. Starting at 0.1, it seems chaotic to start with but after about 70 iterations it suddenly settles to a stable three cycle with values 0.156..., 0.504..., and 0.957....

As we increase μ again, the three cycle becomes unstable at around 3.841 and is replaced by a stable 6-cycle. The period doubling that we saw previously with the two cycle is repeated but rather more frantically with the three cycle. Thus at 3.845 we have a 6-cycle, at 3.848 there is a 12-cycle and at 3.849 there is a 24-cycle but shortly thereafter chaos breaks out again.

There are an infinite number of such stable zones in the generally chaotic area between about 3.57 and 4. Some of the zones are very narrow indeed (they have to be since there are an infinite number of them). There is in fact a zone corresponding to every odd integer and in each zone the period doubling of that integer occurs. Thus there is a zone for 5, 10, 20 and one for 7, 14, 28 and so on.

A 5-cycle zone starts at about 3.739. At $\mu = 3.742$ it has divided into a 10-cycle and at 3.7425 it is a 20-cycle and chaos resumes soon thereafter. There is a 9-cycle at 3.8537 and it becomes an 18-cycle at 3.8539 but 3.854 is chaotic. There is chaos at 3.851492 but a 21-cycle at 3.851493 which becomes a 42-cycle at 3.8515. And so on.

There are some other even cycles as well without a previous odd cycle. Thus there is a 6-cycle zone (but no previous 3-cycle) at about 3.63.

The overall behaviour can be represented by so-called orbit diagrams. The horizontal axis gives the values of μ. The diagram below covers values of μ from 1.0 to 3.54; there is a single stable point of growing value until the doubling

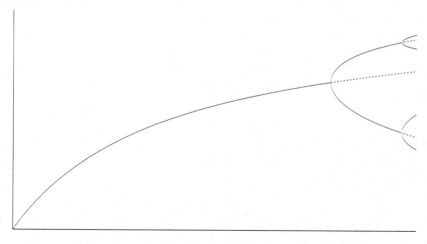

Orbit from $\mu = 1.0$ to 3.54 showing the start of doubling.

8 Chaos and Fractals   191

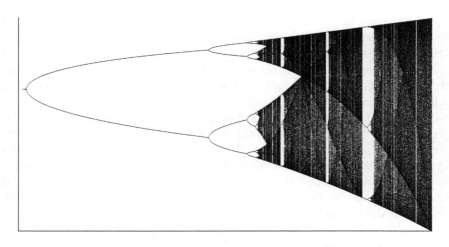

Orbit from μ = 2.99 to 4.0 showing the onset of chaos.

process starts at μ = 3; redoubling occurs at about 3.42. The diagram also shows the unstable fixed points and two cycles as broken lines.

The diagram above covers values from 2.99 to 4 and the lines and dark areas represent the values visited by a process after it has settled down (it thus omits the unstable cycles shown in the first diagram). Note that there are clear regions within the black chaotic area corresponding to the stable zones. The remarkable thing is that if we magnify the area around one of these regions as shown below then we obtain a similar pattern. The whole structure exhibits self-similarity like a fractal.

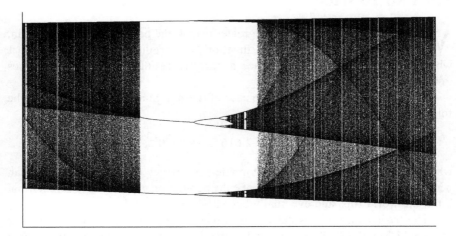

Orbit from μ = 3.8 to 3.9 showing the stable three cycle and its period doubling.

Despite the very strange behaviour, the theory behind this so-called non-linear dynamics was firmly established during the second half of the twentieth century; it was stimulated by the use of computers to explore the topic in a heuristic manner. One notable theorem regarding cycles was discovered by the Russian mathematician N V Sarkovski in 1964. He showed that if a process has a three cycle then it necessarily has cycles of every other value (of course they will almost all be unstable). This seems quite peculiar. It means that since the process for $\mu = 3.83$ has a three cycle then it necessarily has cycles of sizes 28, 209 and every integer imaginable. Of course, we know that it has (unstable) cycles for all the powers of 2 which still lurk from the original doubling process but to have every imaginable cycle is amazing. But this is really just a special case of his theorem. The full theorem says that if we put the integers in the following curious order

$$3, 5, 7, \ldots 2\times3, 2\times5, 2\times7, \ldots 4\times3, 4\times5, 4\times7, \ldots 8\times3, 8\times5, 8\times7, \ldots 8, 4, 2, 1$$

then if a system has a cycle of a particular size N then it has cycles of all sizes following N in the above list. We note that 3 is the first number in the list and so the special case follows.

Another intriguing fact concerns the rate of doubling. Once we are in a doubling sequence, successive doublings occur at an exponential rate. In fact the range of values of μ for each cycle is a factor of about 4.669 smaller than the previous one. This curious constant is known as the Feigenbaum number δ after the American theoretical physicist, Mitchell Feigenbaum, who showed that it appears in many dynamical processes.

And so to dust

WHEN THE VALUE of μ becomes greater than 4, the pattern changes yet again. The key difference is that the function f can produce values greater than 1 which on a further iteration produce a negative result which then progresses towards minus infinity.

As an example, we will take the case of $\mu = 4.5$. Starting with 0.1 we get the following sequence of values

$$0.1,\ 0.405,\ 1.084,\ -0.411\ldots,\ -2.616\ldots,\ -42.570\ldots,\ -8346.836\ldots$$

which illustrates quite clearly that matters soon get out of control. We can perhaps express this by saying that once the population exceeds the 100% normal limit, the world flips and eventually becomes solid antirabbit.

The basic cause of this behaviour is that f is greater than 1 for values of x between 1/3 and 2/3. So once a value lies in the middle third, a catastrophic excursion to minus infinity is inevitable. Of course, if a value becomes exactly

1/3 or 2/3 then the next value is exactly 1 which in turn leads to zero which is an unstable fixed point. So not all points lead to antirabbit, some lead merely to extinction.

Are there any other kinds of points which lead neither to minus infinity nor to zero? Yes there are. For one thing there is the unstable fixed point at $1 - 1/\mu$ which in the case of $\mu = 9/2$ is the point 7/9. Then there are cycles. For example there is the two cycle defined by the equation mentioned earlier

$$\mu^2 x^2 - \mu(\mu+1)x + \mu+1 = 0$$

Inserting $\mu = 9/2$ in this we find that the points of the two cycle are $(11 + \sqrt{33})/18$ and $(11 - \sqrt{33})/18$ which are about 0.930... and 0.291... respectively. This two cycle is unstable. Indeed there are lots of cycles. There are two different three cycles, there are three different four cycles, six different five cycles and so on. All of these cycles are unstable.

The points forming these cycles and the points which lead to zero are in fact some of the points of a Cantor dust with similar properties to the Middle Thirds Set described earlier. Thus it has an infinite number of points but each is quite isolated from other points of the set. The dust consists of all the points which do not eventually lead to minus infinity. We have just seen that the dust includes points which lead to zero and includes points which form endless cycles. Surprisingly, there are also other points which just form endless unrepeating chains of chaos.

In order to get a glimpse of why this might be it is useful to consider the different generations which do not lead to the points which fall into the central third and thus eventually to minus infinity. And we are going to identify the points with a sequence using the letters L and R for left and right. Note that each point X always has two parents (x_1 and x_2) because the equation $f(x) = X$ is quadratic in x and can be written as

$$X = f(x) = \mu x(1-x) \quad \text{or}$$
$$\mu x^2 - \mu x + X = 0$$

and the roots are

$$x = (\mu \pm \sqrt{\mu^2 - 4\mu X}) / 2\mu = \tfrac{1}{2} \pm \tfrac{1}{2}\sqrt{1 - 4X/\mu}$$

Now given any X such that $0 \leq X \leq 1$ these two roots will be real provided $\mu > 4$. Note also that they are symmetrically placed about the central point. Thus in the case $\mu = 4.5$ which we are considering, the two parents of 5/8 are 1/6 and 5/6. That is both 1/6 and 5/6 lead to 5/8. Similarly both 1/9 and 8/9 lead to 4/9.

The two points which lead to 1 are 1/3 and 2/3 and these are denoted by L and R. Similarly the parents of L are denoted by LL and RL and the parents of R by LR and RR. The general rule is that the two parents are denoted by adding L or R to the start of the sequence.

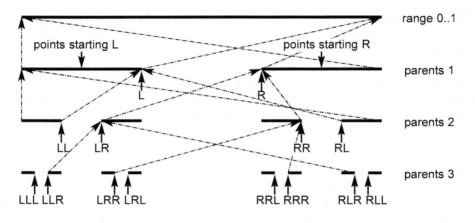

The first few generations of points for μ = 4.5 not leading to death at minus infinity.

This is illustrated by the diagram above. In the first parent generation all the points on the left third lead to a point still in the range 0 to 1; these points have a sequence name starting *L*. Also all the points in the right third also lead to the range 0 to 1 and these have a sequence name starting *R*. Note carefully that the left third maps in the same order but the right third maps in the reverse order. The points *L* and *R* are the extreme points of the two sections as shown and both map to 1. However, all the points in the middle third eventually lead to minus infinity.

We now repeat the process. Thus the parents of the left-hand third are the two outer parts of the thirds, and these points have names starting *LL* and *RL* respectively. The parents of the point *L* are the points *LL* and *RL* themselves. And so it goes on. It should be noted that the process gives rise to a Cantor set very like the Middle Thirds Set but it is not quite the same. *L* and *R* are indeed 1/3 and 2/3, but *LL* is $(9 - \sqrt{57})/18 = 0.0805...$ rather than 1/9 as in the middle thirds set.

Using this technique it is possible to show that the points with names consisting of sequences of letters *L* and *R* correspond exactly to the points of the dust which does not lead to destruction. It is important to note that the points fall into various categories.

There are the points whose name is a finite sequence such as *LRRL*; these points eventually go to 1/3 or 2/3 and so to zero. Note carefully that the successor of a point is obtained simply by removing the first letter, so the points leading on from *LRRL* are simply *RRL*, *RL* and then *L* and *L* is of course 1/3.

But there are also points whose name consists of a recurring sequence such as *LRLRLR*...; these are the points which form cycles. The successor of *LRLRLR*... is *RLRLR*... and its successor is *LRLR*... which of course is where we started so *LRLRLR*... and *RLRLRL*... are the points of the two cycle which we discovered earlier. Note especially that *RRRR*... is the unstable fixed point at 7/9 whereas *LLLL*... is the unstable fixed point at zero.

And finally there are points whose sequence is unending but not repetitive (like the decimal expansion of √2). These are the points which form endless chaotic sequences.

So we have illustrated the behaviour of the points of the dust in terms of what is known as symbolic dynamics. We leave it to the reader to identify the sets of points which give rise to the other cycles such as the three cycles and four cycles and thus to explain why there are several cycles of each order.

It is important to realise that the above discussion only works if μ is greater than or equal to 4. Thus suppose μ is 3 and consider the parents starting at 0.5. They are about 0.212 and 0.788. But if we now compute the parents of 0.788 we find that they are complex numbers and the whole story breaks down.

This completes our exploration of the behaviour of population dynamics. As μ increases further, no other structural changes occur. In summary, when μ is less than 1, the population dies out, when μ is between 1 and 2 it smoothly goes to a stable point, when μ is between 2 and 3 it always overshoots and then oscillates about the stable point. Just above 3, it turns into a two cycle, which divides again and again and then goes chaotic but there are regions of stable cycles until μ reaches 4. Finally when μ exceeds 4 most points diverge away to minus infinity and only a dust of points do not. This dust contains some points which go to zero, some which cycle and some which are endlessly chaotic.

Newton's method

NEWTON DEVISED a simple technique for solving equations. Suppose we have some equation $y = f(x)$ with a graph as shown below. We want to find where it crosses the axis, that is where y is zero. We make a guess and find the gradient at that point and then compute where the tangent crosses the axis and use this as the next guess and keep on repeating until we are satisfied. Sometimes we might have an actual formula for the gradient in which case it is easy to apply.

As a simple example suppose we want to solve $f \equiv x^2 - 9 = 0$. (We know that there are two answers namely 3 and −3.) Using the differential calculus (which Newton invented) the gradient is given by df/dx. Recall the standard formula that the derivative of x^n is $n \times x^{n-1}$. In this case (since the constant term has derivative equal to zero) we therefore find that df/dx is simply $2x$.

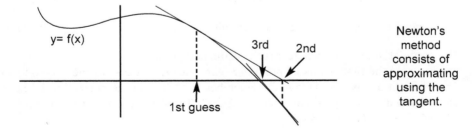

Newton's method consists of approximating using the tangent.

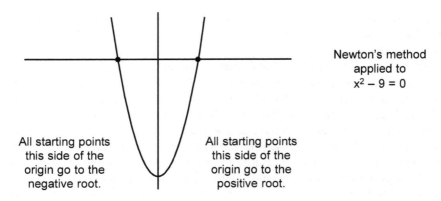

Newton's method applied to $x^2 - 9 = 0$

All starting points this side of the origin go to the negative root.

All starting points this side of the origin go to the positive root.

Let us guess that x is 4 in which case f is $16 - 9 = 7$. The gradient at the point where x is 4 is $2x$ and so has value 8. The adjustment is then error/gradient = $7/8$ = 0.875. So the next guess is $4 - 0.875$ or 3.125. Proceeding in this way we have successive values

$$4.0,\ 3.125,\ 3.0025,\ 3.00000104...$$

and we see that they rapidly approach the right value.

Observe that if we guess with a starting value of x greater than zero then it always goes to the positive answer $+3.0$ whereas if we start with a value less than zero then it always goes to the negative answer -3.0. (Of course if we are dumb enough to start with zero, then the initial gradient is zero and the adjustment gives a division by zero.) So the range of values which go to one answer is neatly divided from the range of values that go to the other answer.

If we work with complex numbers and try to solve $f \equiv z^3 - 8 = 0$ then a strange thing happens. Remember that complex numbers are of the form $x + iy$ where i is the square root of -1. First note that there are three cube roots of 8, the real root $z = 2$ and two complex roots $-1 + \sqrt{3}i$ and $-1 - \sqrt{3}i$. Let us just check the first of these by multiplying it out and remembering that $i \times i = -1$.

$$(-1 + \sqrt{3}i) \times (-1 + \sqrt{3}i) \times (-1 + \sqrt{3}i) = (+1 - \sqrt{3}i - \sqrt{3}i - 3) \times (-1 + \sqrt{3}i)$$
$$= (-2 - 2\sqrt{3}i) \times (-1 + \sqrt{3}i) = (2 - 2\sqrt{3}i + 2\sqrt{3}i + 6) = 8$$

So that's OK.

We can use exactly the same technique as before. The general formula is

next guess of x = (old guess of x) minus (old error f divided by gradient)

The gradient df/dz is $3z^2$ where z is the estimated point.

Suppose the first estimated point is $z = 2i$. The error is $z^3 - 8$ which is $-8 - 8i$. The gradient $3z^2$ is $3 \times (-4) = -12$ (remember $i^2 = -1$). So the correction is the error divided by the gradient which is $2/3 + 2i/3$ and so the next trial point is $-2/3 + 4i/3 = -0.666... + 1.333...i$.

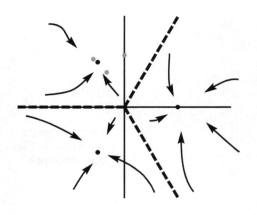

Expected three regions of the complex plane which one might think would contain starting points converging on the three cube roots of 8.

Proceeding in this way we get a sequence of values (to three decimal places) as follows

$$2i, \; -0.667+1.333i, \; -1.164+1.848i, \; -1.017+1.736i, \; -1.000+1.732i$$

And, as we see, they converge onto $-1.000 + 1.732...i$ which is the correct result. The first three are shown on the diagram above in red, the roots are shown in black. Now since there are three roots and these are neatly placed at 120° around the origin (zero) we might expect the regions containing the starting points which lead to the individual roots to neatly divide the complex plane into the three triangular lumps as shown in the same diagram (basins of attraction is the technical name).

But the three regions are not like that at all. Consider as another example starting at $1 + i$. We might expect that this would converge to the real root at 2 because that is nearest. But it doesn't. The diagram below shows the erratic initial behaviour. The first correction leaves the estimate nearest to the real root but the next correction moves towards the root in the upper half of the plane to which it eventually converges.

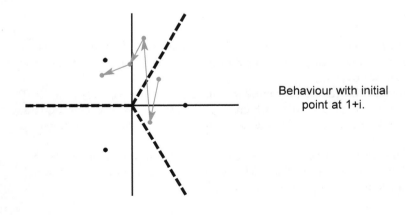

Behaviour with initial point at 1+i.

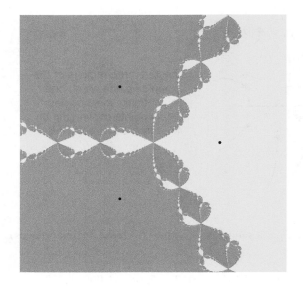

The three basins of attraction to the cube roots of 8.

In fact the basins of attraction are very strange with fractal boundaries with the curious property that each boundary point is in fact a boundary point of all three regions. As a consequence there are starting points leading to each root from far within the other regions. The diagram above shows the three regions in different colours with the roots marked in black.

Julia and Mandelbrot sets

WE HAVE DISCUSSED at length the population equation $f(x) = \mu x(1-x)$ and showed that it had different behaviour for various values of μ. The key point about the equation is that it is non-linear – that is it contains powers of x other than just x itself. A related quadratic equation which we will now look at is

$$Q(x) = x^2 + c$$

This has similar behaviour to f for real values of x where the parameter c takes the place of μ although as μ increases c decreases and so the orbit diagram is reversed as shown opposite.

The general behaviour is very similar. Thus $c = 0.25$ corresponds to $\mu = 1.0$. The two cycle starts at $c = -0.75$ which corresponds to $\mu = 3$. Chaos starts at around $c = -1.4$ and complete divergence occurs at $c = -2$ which corresponds to $\mu = 4$.

Within the chaotic zone from -1.4 to -2 we find zones of stable cycles and period doubling just as before. Thus there is a three cycle at around -1.75 which doubles to a six cycle at around -1.77.

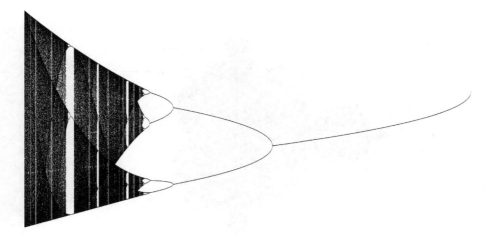

The orbit diagram for the function $Q(X) = x^2 + c$ with c from -2.0 to $+0.25$.

The function Q can of course also be applied to complex numbers z and we then write

$$Q(z) = z^2 + c$$

where c may also be complex.

Gaston Julia (1893–1978) and Pierre Fatou (1878–1929), both French mathematicians, studied the behaviour of dynamical systems in the early part of the twentieth century. There are two types of sets known as Julia sets. The filled Julia set for a function such as Q is defined as the set of points z which do not cause z to iterate off to infinity. The Julia set is the boundary of the filled Julia set.

In the case of the function f with μ greater than 4 we saw that the set of points which did not go to minus infinity was a Cantor set. So the Julia set for that example is that Cantor set. Moreover, since the Cantor set was simply a set of isolated points, it follows that the filled Julia set and the Julia set are the same in that case.

In the case of the function Q we find that for some values of c, the (filled) Julia set covers a solid region. Thus if c is zero then the function Q becomes

$$Q(z) = z^2$$

and so it is simply a squaring function. In the complex plane if z is in the unit circle then the points converge on zero whereas if it is outside then they go to infinity. The points on the unit circle stay on the unit circle. Some points of the unit circle exhibit cycles and some are chaotic. The filled Julia set is itself the unit circle.

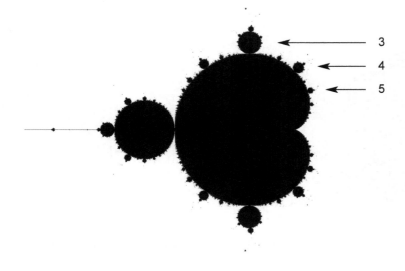

The Mandelbrot set.

On the other hand if we take c to be quite large such as 10 then all points go to infinity and the Julia set is empty. For some values of c, the Julia set consists of a disconnected dust whereas for others it consists of a set of branching lines often called a dendrite.

The Mandelbrot set is a sort of key to the Julia sets. It is named after Benoit Mandelbrot who discovered it in 1980. Again it is a map in the complex plane. A point c is in the Mandelbrot set if the filled Julia set for c is connected (that is not just a dust). The Mandelbrot set is another example of a fractal with self-similar portions. A simple image of the set is shown above. It is important to realise that it is difficult to see what the Mandelbrot set really looks like. This is because it has dendritic arms sprouting from the various lumps and these are curiously shaped lines of zero width. And since a line has zero width, it is invisible. Within the arms are solid regions which are part of the set and these show as spots in the diagram. Amazingly, these small solid regions have a similar shape to the whole set thus showing self-similarity once more.

It can also be shown that the Mandelbrot set is the set of points c for which

$$Q(z) = z^2 + c$$

does not go to infinity when starting at $z = 0$.

With $c = 0$, the sequence of values of Q starting with $z = 0$ is 0, 0, 0, ... and so we have a fixed point at the origin and 0 is in the Mandelbrot set.

With $c = 1$, the sequence commences 0, 1, 2, 5, 26, and clearly this diverges to infinity and so 1 is not in the Mandelbrot set.

With $c = -1$, the sequence is 0, −1, 0, −1, ... and this is a two cycle and so −1 is in the Mandelbrot set.

In fact, the main cardiod-shaped zone is that for which there is a fixed point. The circular zone to the left of that is a two-cycle zone and the smaller circular zone to the left of that is a four-cycle zone and so on thus exhibiting period doubling. On the negative axis it is then chaotic but does not diverge to infinity. Moreover, a tiny version of the Mandelbrot set can be seen on the negative axis and this covers the stable three cycle and its doublings inside the chaotic region.

The various bulbs around the main zones correspond to other stable periods. For example, the large bulbs attached to the top and bottom of the central cardiod are zones of three cycles (such as $c = -1/8 + 3/4i$). The bulbs at about SE and NE are zones of four cycles (such as $c = 1/4 + 1/2i$) and the next significant bulbs are zones of five cycles (such as $c = 3/8 + 1/3i$).

Another representation of the set is shown below. The green regions are points near the set. They are in fact points which have not diverged beyond a circle of radius 5 after some 20 iterations of the function Q. They illustrate the general position of the dendritic arms.

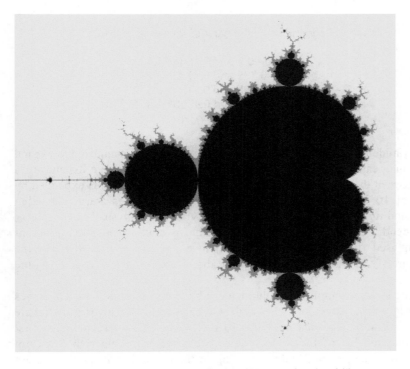

The Mandelbrot set in colour hinting at the dendritic arms.

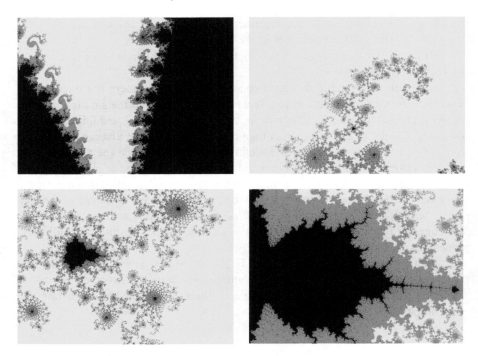

The four diagrams show the parts of the set from about

−0.8 to −0.7 and 0.15 to 0.25
−0.745 to −0.735 and 0.17 to 0.18
−0.7405 to −0.7395 and 0.172 to 0.173
−0.7402 to −0.7401 and 0.17255 to 0.17265.

The Mandelbrot set exhibits amazing complexity and it is amusing to expand a region in greater and greater detail. The four diagrams above expand on a region between the main cardiod and the period two circle. Each is expanded by a factor of 10 on the previous one. The Mandelbrot set is a truly amazing structure and turns up in the analysis of stability in many physical processes.

The dendritic arms themselvers are usually invisible because the points of the bitmap used to create the diagrams rarely lie exactly on an arm. However, the chaotic part of the real axis shows because the bitmap points lie exactly on the axis.

An interesting example of a point that can be seen is that for $c = i$. This gives a two cycle whose values are $-1+i$ and $-i$. Similarly $c = -i$ gives a two cycle whose values are $-1-i$ and i. These two points can just be seen in the first coloured diagram. That for i is in red and will be found near to the right hand arm sprouting from the period 3 bulb at the top of the cardiod. A magnifying glass will probably be required to find it.

Natural chaos

WE HAVE INTRODUCED chaos through the example of the growth of the rabbit population. But chaos abounds everywhere. Most physical processes are governed by non-linear equations and so exhibit chaos.

In fact it is only a few processes that are linear and so are easily predicted. The oscillation of a pendulum is one where a slight disturbance of the initial conditions does not result in wildly different behaviour. The behaviour of a single planet around the sun is also a linear process and so easily predicted. But the general three body problem (such as earth, moon and sun) is not so easy and has to be done by approximations (although these are in this case quite accurate).

Good examples of non-linear systems which affect us greatly occur in fluid dynamics which describe the movement of water and air and thus the sea and the weather. The main equation concerned is the Navier–Stokes equation which is simple to express but is not linear. No matter how good our knowledge of the current values of the pressure and temperature, wind speed, etc. throughout the world it is really impossible to predict the weather for more than a few days ahead. The smallest difference soon gives rise to widely different behaviour and hence the fable of how killing a butterfly in England can cause a storm in Singapore.

The movement of the sea bashing against rocks is almost all simply the Navier–Stokes equation. But what a mess; one could not conceive of predicting just how the sea will behave down to the last drop. The sea is a splendid example of chaos in action.

Further reading

THE BOOK *Chaos* by James Gleick is a partly historical overview of the subject with pretty pictures and no mathematics. Two books with excellent illustrations are *The Beauty of Fractals* by Peitgen and Richter and the *Science of Fractal Images* which is a collection of essays by Barnsley, Devaney, Mandelbrot, Peitgen, Saupe and Voss. These books both have hard lumps of mathematics but they can be appreciated and dipped into without too much trouble.

For a proper understanding of population dynamics and the theory behind Julia and Mandelbrot sets, I can thoroughly recommend either *A First Course in Chaotic Dynamical Systems* or *An Introduction to Chaotic Dynamical Systems* both by Robert Devaney. *A First Course* is easier and was written later. *An Introduction* was written earlier and goes further. Both are undergraduate texts and should be accessible to anyone who enjoyed mathematics at school (and is diligent). Both describe period doubling, symbolic dynamics and so on. *A First Course* mostly uses the function $Q = x^2 + c$ whereas *An Introduction* uses $f = \mu x(1-x)$ more as in these notes.

Exercises

1. a) What is the dimensionality of the curve formed by repeatedly subdividing the line fragment below with reduced versions of itself as described in the text?

   b) A square has side of length three units and is subdivided into nine squares of side 1. The centre square is replaced by a cube of side 1 (strictly by a cube with the base missing). Each unit square of the figure is then replaced by a copy of the whole figure with one third the linear dimension. This replacement is repeated ad infinitum. What is the dimensionality of the resulting surface?

2. In the population growth example, suppose $\mu = 3^4/_{15}$. The population eventually settles into a two cycle. What are the two values of the cycle and what is the value of the (unstable) fixed point?

3. In symbolic dynamics, there is one two cycle namely *LRLRLR*.... Explain why there are two three cycles and give their sequences. Similarly give the sequences for the four cycles and five cycles.

Line segment for 1a.

9 Relativity

EINSTEIN PROPOSED two theories of relativity at the start of the twentieth century which painted a picture of the world quite at odds with our intuition regarding the nature of space and time.

The two theories are known as the Special Theory and the General Theory. The Special Theory essentially concerns the velocity of light and the mathematics is fairly easy to understand even though the outcomes are quite startling. The General Theory concerns gravitation and the mathematics behind it is considered rather difficult (postgraduate in these dumbed-down days I am sure). The effects of both theories differ little from the traditional predictions of Newton except in extreme circumstances. Nevertheless, in those extreme circumstances all experiments have indicated that they are correct.

This lecture gives a reasonable presentation of the Special Theory and explains phenomenon such as time and space contraction. It also gives just a glimpse of the General Theory which lies behind such things as Black Holes.

The special theory

THE SPECIAL THEORY arose out of the problem of the velocity of light. Scientists were familiar with the idea that waves represented a disturbance in some medium. Thus sound is propagated as a disturbance in the air and the waves on the sea are obviously disturbances of the surface of the water. It was known that light was a form of electromagnetic wave and it seemed natural to assume that it was a disturbance in something. Since light reaches us from the Sun and stars this stuff clearly had to be pervasive. It was called "the ether" or often "the æther" in order to distinguish it from the common chemical $C_2H_5.O.C_2H_5$ (properly called diethyl ether).

Many attempts were made to find how fast the sun and earth were travelling through the ether. If the velocity of light in the ether were c and the ether were travelling relative to the earth at a velocity e then we would expect that the measured velocity of light relative to the earth would vary between $c-e$ and $c+e$ according to the direction of measurement. Of course light is pretty speedy at some 300,000 km per sec and so the variation might be hard to detect. But light is not that fast – it takes 8 minutes for the light from the sun to reach us and the delay in radio communication over long terrestrial distances, especially if satellites are involved, is noticeable.

The famous experiments of Michelson and Morley in 1887 showed that the velocity of light was always the same no matter how they measured it. They measured the time taken for light to go from a source to a mirror and back and did it in two directions at right angles. The time was the same although it shouldn't have been by traditional Newtonian mechanics. It was tried when the earth was in different positions in its orbit around the sun thereby eliminating the possibility that the earth was coincidentally at rest relative to the ether at the time of the experiment.

An example which can be used to illustrate the expected effect is that of a duck crossing a river. Suppose she can paddle at 5 feet per second and that the river is 50 feet wide. Then to go across and back is clearly 100 feet and so takes 20 seconds provided there is no current. Equally to paddle along the river for 50 feet and then back again will also take 20 seconds if there is no current.

Now suppose there is a current of 3 feet per second. Then when going across, the duck has to aim slightly upstream in order to avoid being swept downstream. The result is that because of the composition of vectors using Pythagoras, the actual net speed across is only $\sqrt{(25-9)} = \sqrt{16} = 4$ feet per second and so the time taken for the return trip is 25 seconds.

The trip up and down for 50 feet takes rather longer. When paddling downstream the duck is helped by the current and so goes $5+3 = 8$ feet per second whereas when going upstream, the current is against her so that she only goes $5-3 = 2$ feet per second. So to do 50 feet with the current takes only $50 \div 8 = 6.25$ seconds but 50 feet against the current takes $50 \div 2 = 25$ seconds. Therefore the return trip takes 31.25 seconds which is quite a lot more than the time for the double crossing.

The experiments were checked and double-checked and the conclusion was that the velocity of light was always the same no matter how it was measured. A number of hypotheses were proposed in order to explain this result. The Irish physicist Fitzgerald (1851–1901) and the Dutch physicist Lorentz (1853–1928) both proposed that bodies moving through the ether suffered a contraction along their length. This certainly explained the Michelson and Morley results but was essentially tampering with the symptoms rather than getting to the disease which caused them. Other proposals concerned the effect of the ether on moving clocks.

In 1905, Albert Einstein proposed the radical solution that there is no ether at all and the velocity of light (and indeed the whole of physics) is the same *relative* to all observers in uniform motion no matter how they move. This also dispensed with the common sense idea that distances and time intervals are the same for all observers. This is a big traumatic shock. We are so used to the idea of absolute time that it is very hard to get used to the idea that time depends upon by whom it is measured.

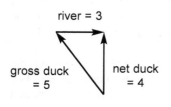
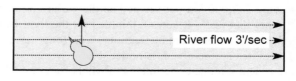

Composition of vectors for duck crossing river.

Time changes

IN ORDER to contemplate the effect of relativity on time we can carry out some thought experiments. Imagine a VHST (very high speed train) travelling at a velocity v. Inside the train is a sort of light clock which bounces photons back and forth between two mirrors. Suppose the distance between the mirrors is d. If the velocity of light is c then clearly the time for the light to go from one mirror to the other as measured by the conductor on the train is d/c. We will call this time interval t'. So we have $d = c \times t'$.

But for the train spotter on the ground the effect is quite different. By the special theory the velocity of light as measured by him is still c. But in a time interval t for the train spotter in which the light goes from one mirror to the other, the train will have moved a distance vt and so we have a triangle in which the hypotenuse is ct and the base is vt. Note that, for the observer, the photons bouncing between the mirrors are travelling at an angle and so traverse the hypotenuse. So we must have (by Pythagoras)

$$c^2 t^2 = c^2 t'^2 + v^2 t^2 \quad \text{or}$$

$$t' = t \times \sqrt{(1 - v^2/c^2)}$$

So the time as measured by the observer on the train is less than that measured by the train spotter. In other words the clock on the train seems to be (nay, actually is) slowed down according to the train spotter by the factor $\sqrt{(1 - v^2/c^2)}$.

Since c is quite large (300,000,000 metres/sec), the effect is fairly small for normal trains. For a train travelling at 360 km/hour ($v = 100$ metres/sec), v/c is only 0.00000333.... But for a supergalactic train travelling at half the velocity of light, the effect is significant since then v/c is 0.5 and $\sqrt{(1 - v^2/c^2)}$ is 0.866....

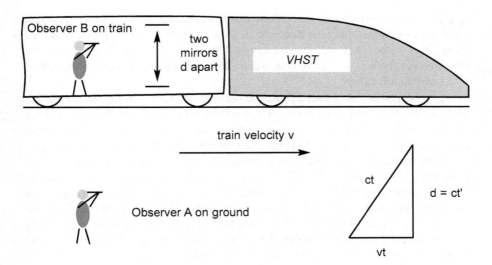

Such enormous speeds seem unlikely but the effect is observed for particles reaching the earth from outer space. Certain radioactive particles generated at high altitude reach the surface of the earth in a proportion which is unexpectedly high. This is because they decay with a half life as measured by their internal clock. Consider particles travelling at $0.866c$ towards the earth; this velocity has a time factor change of 0.5. Suppose that the half life of decay of the particles (as measured when stationary in the laboratory) is 0.001 sec and we observe that the particles take 0.002 sec to reach the earth as measured by us. We would expect only one quarter of the particles to reach the earth. But, because of the time effect, the time as measured by the internal atomic activity of the particles is only 0.001 sec and so half of them actually reach the earth.

There are well known "paradoxes" concerning a pair of twins, one who stays at home and the other who goes on a trip to a star and back. Suppose that the mobile twin travels at 99% of the velocity of light. Then his clock will appear to be running at about 1/7th of its natural rate according to his sister who stays at home. A round trip to a star 25 light years away will take 50 years as measured by the sister but only 7 years as measured by the traveller. This is genuine – when they meet again, the sister will have a blue rinse but the traveller will still be a young man.

It is often thought that this is a paradox because surely according to the traveller, the sister's clock should have been running slowly and so she should seem young to him. But the situation is not symmetric since the experience of the traveller is quite different in one important respect – he suffered enormous accelerations and these have to be taken into account in a full analysis.

The Lorentz–Fitzgerald contraction

WE MENTIONED earlier that Lorentz and Fitzgerald both hypothesized that bodies shrink along the direction of their travel through the ether. Although there is no ether and so the framework for their hypothesis is irrelevant, nevertheless there is a relative contraction in the direction of motion and this is often named after them.

Consider once more the radioactive particles on their journey to the earth's surface. We noted that they decayed at a different rate to that expected from the point of view of the observer on earth and we concluded that this was because of time going more slowly on the particle as measured by the observer on earth.

Now consider a (small) observer riding with the particles. A key point about the earth view was that only half of the particles decayed before reaching earth. But this must also hold from the particles' view because 1/2 is simply a raw numerical proportion and not a physical constant. The time of travel from the particles' view thus must be 0.001 sec (the half life). However, the velocity being relative must be the same whether the particle measures the velocity of the earth coming towards it or vice versa. Well, the overall story must be consistent from both viewpoints and the only conclusion has to be that from the particles' viewpoint the distance to the earth as measured by the particle must be only one half of the distance as measured from the earth. So the earth-measured distance contracts for the particle by the same factor as the particle-measured time contracts for the earth.

This is known as the Lorentz–Fitzgerald contraction – an object whose length is L when at rest (as measured by an observer moving with the object) has length

$$L' = L \times \sqrt{(1 - v^2/c^2)} \qquad \text{the Lorentz–Fitzgerald contraction}$$

as measured by an observer travelling at velocity v relative to the object along its length. (Note carefully that the width of the object is the same to both observers and does not change.)

There are some amusing examples mentioned by Rindler. A man wishes to store a 40 foot pole in a 20 foot garage. He runs into it at $0.866c$ and so to an observer in the garage it shrinks to 20 feet and exactly fits and the observer quickly closes the door. We imagine a large concrete block at the end of the garage which prevents the pole from bursting through the wall. But once the pole is stopped it becomes 40 feet long again and breaks through the door!

Gosh you might say but what about the viewpoint of the pole? It finds it is approached by a garage which is only 10 feet long so how on earth can it fit?

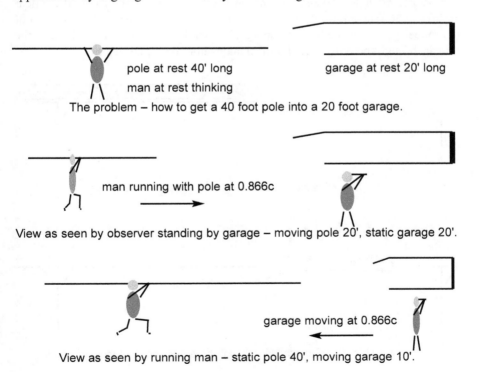

pole at rest 40' long
man at rest thinking
The problem – how to get a 40 foot pole into a 20 foot garage.

man running with pole at 0.866c
View as seen by observer standing by garage – moving pole 20', static garage 20'.

garage moving at 0.866c
View as seen by running man – static pole 40', moving garage 10'.

Well, the garage keeps going when it hits the pole and takes the front end of the pole with it. The shock wave down the pole cannot travel faster than c and so the back end doesn't know about the impact for some time and remains stationary until the garage engulfs it. Note that from the viewpoint of an observer who was travelling with the pole but escapes being engulfed by the garage, the 40 foot pole is engulfed by a 10 foot garage which carries the pole along with it and squeezes it to fit on impact. But (as seen by the observer who was with the pole and continues in motion relative to the garage) once the pole is in uniform motion again and being carried along by the garage, the original 40 foot pole has a contraction to 20 feet and so is still twice as long as the 10 foot garage and thus bursts through the door. We get the same overall story however we look at it.

Distortion of bodies

WE HAVE JUST OBSERVED that bodies shrink in motion. But it is very much more curious than that.

Consider a railcar of length 60 feet. There is a hole in a bridge of length 40 feet and a very strong wind blowing in gusts. The driver decides to make sure that he can get the railcar over the hole by travelling at high speed and in fact approaches the hole at $0.866c$. From the point of view of the bridge, the railcar shrinks to 30 feet and easily fits into the hole. Moreover, just as the back end of the railcar gets over the hole, a supergiant instantaneous gust of wind blows the railcar through the hole while it still remains horizontal. Disaster!

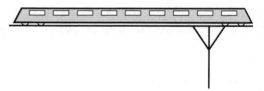

Static view of railcar 60 feet long and 40 feet hole in bridge.

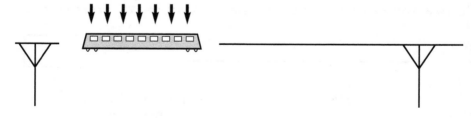

Gust of wind blowing short railcar through hole from point of view of bridge.

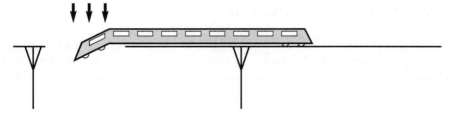

Situation from point of view of railcar when gust is only acting on front of car.

But now look at it from the point of view of the railcar. The hole in the bridge is now only 20 feet long and it looks as if the rigid railcar will easily get over the hole. What happens? The key is that simultaneous events are only simultaneous at the same time *and place*; events at the same time but different places for one observer will be at different times for a moving observer. So the gust of wind hits the different parts of the railcar at different times from the point of view of the railcar. In fact it hits the front of the railcar well before it reaches the opposite side of the hole; the front of the railcar bends down and the whole thing wiggles through the hole.

So we have to come to the very surprising conclusion that there is no such thing as a rigid body.

In *Mr Tompkins in Wonderland*, George Gamow describes the experiences of Mr Tompkins in a world where the velocity of light is rather small, perhaps 30 mph. He describes how he sees a cyclist whose wheels have shrunk in the direction of motion and thus become ellipses. And then when he ventures forth on a bicycle he finds that the city blocks all shrink around him.

But the perceived distortion is more complex than that. Moving objects actually become curved from the point of view of the observer. Rectangular city blocks will lean over. The basic reason is that the image we see is produced by photons all arriving at our eye at the same time. This means that we see parts of an object that are further away as they were at an earlier time than those parts which are nearer to us. (Of course this would apply without relativity, it is simply a consequence of the finite velocity of light, but the effect is strangely modified by relativity.)

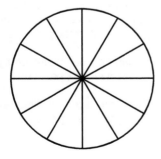 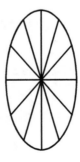

A wheel shrunk by the Lorentz–Fitzgerald contraction.

Lorentz transformation

IN THE GENERAL CASE the situation as seen by an observer O travelling at velocity v relates to the situation seen by another observer O' according to the following equations – we assume that the axes have the same orientation and that the motion is along the x-axis

$$x' = \gamma(x - vt), \quad y' = y, \quad z' = z, \quad t' = \gamma(t - vx/c^2) \quad \text{where } \gamma = 1/\sqrt{(1 - v^2/c^2)}$$

These give the positions and time as seen by one observer in terms of those seen by the other. Note that the y and z values are the same. The x transformation involves the time as expected but also involves the factor γ. The t transformation shows that the times as measured by the two observers are different and moreover reveals the factor γ as well corresponding to the fact that the clock rates differ as measured by each other.

The fact that both the x and t transformations involve both x and t implies that time and space are mixed up. However, although distances and times as measured by the two observers of two different events E_1 and E_2 are different, nevertheless the observers will agree on the value of

$$\delta s^2 = \delta x^2 + \delta y^2 + \delta z^2 - c^2 \delta t^2 \qquad \delta s \text{ is the interval between two events}$$

where δx, δy, δz are the differences between the positions of the two events and δt is the difference between the time of two events. The value δs is known as the interval between the events and is the same for all observers.

Note that δs^2 is zero for two events on a beam of light. If it is negative then one event can communicate with the other by sending a signal; if it is positive then neither can communicate with the other and happenings at one event cannot influence the other.

This is traditionally depicted by a so-called light cone as shown below. We can suppress one of the space dimensions (z, say) and so we are effectively considering relativistic Flatland.

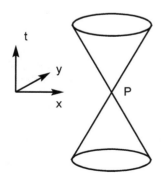

The future

The present

The past

The event at P can only influence events within the future cone.

It can only be influenced by events in the past cone.

The cone traces possible rays of light passing through P.

Time and relativity

ONE OF THE PROBLEMS with dealing with relativity is accepting the fact that events that are simultaneous for one observer will not be simultaneous for an observer moving relative to the first one.

Consider the problem of a train 500 metres long that makes marks on the track at each end of the train. The train has a central switch which sends a signal to the two ends in an identical manner and these then trigger the track marking devices. In an identical manner means they travel to the two ends of the train at the same speed and therefore the marks are made on the track at the same time according to an observer on the train.

Since the ground is contracted from the viewpoint of the observer on the train by the Lorentz–Fitzgerald contraction it means that the marks on the ground are more than 500 metres apart. This sounds wrong because to an observer on the ground the train will have contracted and it is natural to assume that the marks are therefore closer apart than 500 metres. The solution to the dilemma is that from the viewpoint of the observer on the ground the train does not make the marks at the same time. It makes the mark at the rear of the train first and then later makes the mark at the front and meanwhile of course the train has moved on so that the marks end up further apart than expected.

Let us quantify the effect. We will assume that the train is travelling at $0.6c$ at which speed the contraction is 0.8; this gives convenient numbers. According to the train the marks will therefore be $500/0.8 = 625$ metres apart.

Suppose that the signals are electromagnetic and therefore travel to the ends of the train at the velocity of light c. They will travel at this velocity according to the observer on the ground as well because that is the nature of light and relativity.

Suppose the centre of the train is at coordinate $x = 0$ as measured on the ground at the time $t = 0$ according to the ground clock when the signal is sent to the two ends. Suppose the train is going in the direction of increasing x. At time $t = 0$ the front of the train will be at $x = 200$ metres and the back will be at $x = -200$ metres. Remember that from the ground the train will have contracted to $500 \times 0.8 = 400$ metres.

Consider the front of the train first. At time $t = T_f$, it will be at the point $x = 200 + 0.6cT_f$. The signal which is travelling at velocity c will at that time be at the point cT_f. So

$x = 200 + 0.6cT_f$ position of front of train at time T_f

$x = cT_f$ position of signal at time T_f

Now when the signal reaches the front of the train clearly these two values for x must be the same since both signal and front of the train will also be at the same place as seen by the observer on the ground. By equating these we can find the time on the ground clock at which the mark at the front is made. We have

$200 + 0.6cT_f = cT_f$ which gives $0.4cT_f = 200$ so that

$T_f = 500/c$ time of front mark

We can now do the same for the back of the train. At time T_b the back of the train will be at the point $x = -200+0.6cT_b$ and the signal (which is travelling in the opposite direction) will be at $-cT_b$. Equating these gives the time on the ground clock when the mark at the back is made. We have

$-200 + 0.6cT_b = -cT_b$ which gives $1.6cT_b = 200$ so that

$T_b = 125/c$ time of back mark

We now see clearly that the marks are made at different times.

We can now compute the position of the marks by putting the values of the times into the formulae for the positions of the signals. We obtain

front mark at $x = cT_f = 500$ using position of signal

back mark at $x = -cT_b = -125$ using position of signal

We can check this by putting the times into the formulae for the position of the ends of the train thus

front mark at $x = 200+0.6cT_f = 200 + 0.6c \times 500/c = 200+300 = 500$

back mark at $x = -200+0.6cT_b = -200 + 0.6c \times 125/c = -200+75 = -125$

The distance between the marks is therefore $500 - (-125) = 625$ which is the same as deduced by the observer on the train.

So the two views do give the same answer but clearly it is much easier to choose the observer for whom the events are simultaneous.

This can be illustrated by the diagram opposite which plots the position of the ends of the train and the signals against time as measured by the observer on the ground. The two signals start at the origin and since they travel at the velocity of light, their paths are a section of the light cone from that point.

The calculation is much more tedious if the signals going to the ends of the train do not travel at the speed of light but at some lesser speed v. The difficulty is that the speed of the two signals as seen by the observer on the ground are not v and indeed are not the same. One way to do the calculation is to use the formula for the composition of relative velocities. If an object A is travelling at velocity v relative to the observer and B is travelling at velocity u relative to A then the velocity of B relative to the observer is

$(v + w) / (1 + vw/c^2)$ composition of velocities

Note that if v and w are both very small so that vw/c^2 is negligible then the formula becomes $v + w$ as we expect. Also if one of them, say w, equals the velocity of light c then the result is $(v + c) / (1 + v/c) = c$ again as we would expect.

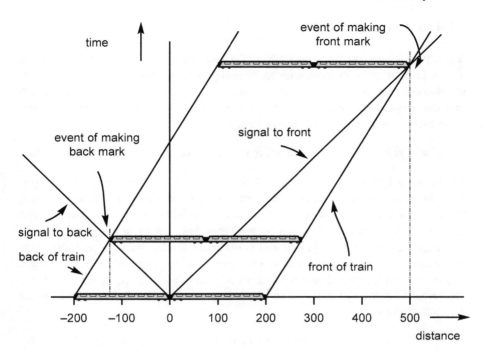

We can now do the general calculation. Suppose the train is travelling at velocity v and the signals down the train travel at velocity w. Then the signal going to the front of the train as seen by the observer on the ground has velocity $(v+w)/(1+vw/c^2)$ and the signal to the back has velocity $(v-w)/(1-vw/c^2)$. For convenience let us put $\gamma = \sqrt{(1-v^2/c^2)}$, the contraction factor. At time $t = 0$ the ends of the train are at $x = +250\gamma$ and $x = -250\gamma$. As before we compute the times T_f and T_b at which the marks are made by equating the positions of the ends of the train and the signals.

$$250\gamma + vT_f = (v+w)T_f / (1+vw/c^2)$$
$$T_f = 250\gamma(1+vw/c^2)/(w(1-v^2/c^2)) = 250(1+vw/c^2)/w\gamma$$
$$-250\gamma + vT_b = (v-w)T_b / (1-vw/c^2)$$
$$T_b = 250\gamma(1-vw/c^2)/(w(1-v^2/c^2)) = 250(1-vw/c^2)/w\gamma$$

The positions of the marks are

front mark at $x = (v+w)T_f / (1+vw/c^2) = 250(v+w)/w\gamma$

back mark at $x = (v-w)T_b / (1-vw/c^2) = 250(v-w)/w\gamma$

and the difference between these is simply $500/\gamma$ which is the length of the train divided by the contraction factor and so is greater than 500 exactly as computed by the observer on the train.

Mass and energy

IN NEWTON'S LAWS, the acceleration of a body depends upon the force applied to it and the velocity continues to increase smoothly without limit. However, in relativity this cannot be so because the velocity cannot exceed c. It turns out that the mass of the body is not constant but increases with velocity using the same factor $\gamma = 1/\sqrt{(1 - v^2/c^2)}$. So we have

$m = m_0/\sqrt{(1 - v^2/c^2)}$ mass is rest mass multiplied by the magic factor

So the mass keeps on increasing and becomes infinite at the velocity of light. This increase in mass absorbs the energy and prevents the velocity from increasing without bounds. This means that material bodies cannot travel at the velocity of light. But photons can because they have zero rest mass.

This increase in mass is real and is observed for fast moving particles such as electrons. An electron moving in a magnetic field experiences a force at right angles to its velocity and as a result goes around in a circle. If the field is H and the velocity of the electron is v and its charge is e then the force caused by the magnetic field is

$F = Hve/c$ force on charged particle moving in magnetic field H

This is counterbalanced by the centrifugal force which is

$F = mv^2/r$ centrifugal force

where m is the mass of the electron and r is the radius of the circle. These two expressions for F must balance and so we can deduce that the radius must be

$r = mvc/He$ radius of orbit of electron

and so the faster the electron goes the bigger the circle it describes. Since the circumference of the circle is $2\pi r$ it follows that the time per revolution is

$T = 2\pi r/v = 2\pi mc/He$ time per revolution

which is independent of the velocity. This is the principle of the cyclotron, a classical device for accelerating particles whereby they are given a pulse of energy each cycle and move into a wider orbit. However, as the electrons speed up the value of m increases and the time of revolution changes slightly. This meant that the simple cyclotron could not be used to generate very fast electrons and had to be modified to form the synchrotron in which the timing of the pulses of energy is synchronized to the circular beam of ever more massive electrons.

Let us now consider the momentum and energy of moving bodies. Remember that momentum (mv) is conserved but that energy ($\frac{1}{2}mv^2$) can turn

9 Relativity

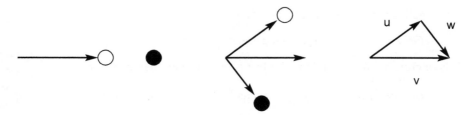

If a billiard ball hits one that is stationary, they move apart at right angles according to Newton.

into heat. When two cars collide much of the energy of their motion is converted to heat, some to sound and so on. But the total momentum is never changed.

An interesting example concerns billiards. If a ball hits a similar stationary ball at an angle then both balls end up moving. Suppose the velocity of the one ball beforehand is v and afterwards is u and the other ball has velocity w afterwards. The velocities fit together in a triangle as shown above in order to conserve momentum. However, if no energy is lost we also have

$$\tfrac{1}{2}mv^2 = \tfrac{1}{2}mu^2 + \tfrac{1}{2}mw^2 \qquad \text{conservation of energy}$$

So by Pythagoras the triangle has a right angle and so the balls move apart along lines at right angles as shown above. All billiard players know this. But this is not true in relativistic billiards so Mr Tompkins would be in difficulties if he ventured into the billiard saloon.

We have seen that distance and time are confused in relativity. Consider a uniform force F acting on a body in the Newtonian world. The change in momentum of the body is F multiplied by the *time* for which the force is applied whereas the change in energy is F multiplied by the *distance* the body moves while the force is applied. Since distance and time are confused, it is not surprising then that momentum and energy are confused in relativity as well. Instead of talking about momentum and energy separately in fact we talk about the energy–momentum vector which has four components. Three (corresponding to the three space dimensions, x, y, z) give the momentum and the fourth (corresponding to time t) is the energy.

In order to get the transformations to work if we move to a different observer we find that we have to have

$$P = (mv_x, mv_y, mv_z, m) \qquad \text{energy–momentum vector}$$

So the mass is identified with the energy. Moreover, we find that if a force F adds an amount of energy dE to a moving body then we find that the increase in velocity is precisely that corresponding to an increase in mass dm given by

$$dE = dm \times c^2 \qquad \text{increase in energy relates to increase in mass}$$

This implies that the whole energy of a body is given by

$$E = mc^2 \qquad \text{the famous relation between mass and energy}$$

It is interesting to write this in the form using m_0 and then to expand it

$$E = m_0 c^2 / \sqrt{(1 - v^2/c^2)} \qquad \text{expand using binomial theorem}$$
$$E = m_0 c^2 + \tfrac{1}{2} m v^2 + \ldots$$

and this reveals that the total energy is the rest energy plus the normal Newtonian kinetic energy $\tfrac{1}{2}mv^2$ to a first approximation.

That electricity and magnetism are mixed up was known for many years before Einstein. A moving magnet creates an electric current and a moving charge (an electric current) creates a magnetic field. Relativity greatly simplified Maxwell's equations. In fact they are best discussed in terms of the so-called electromagnetic tensor. A tensor is like a vector but has two dimensions of components. So in four-dimensional space–time a tensor has 16 entries. That describing electromagnetism can be laid out as follows

$$F_{ij} = \begin{pmatrix} 0 & H_3 & -H_2 & -E_1 \\ -H_3 & 0 & H_1 & -E_2 \\ H_2 & -H_1 & 0 & -E_3 \\ E_1 & E_2 & E_3 & 0 \end{pmatrix}$$

The Es are the electric field and the Hs are the magnetic field in the different dimensions. The rate of change of F is in fact the current or charge. Note that the Es and Hs are involved in different ways. This relates to the fact that there are no free magnetic poles.

If we have matter distributed around rather than as single point objects then it turns out that energy and momentum also have to be described by a tensor. In fact both the mechanical and electric energy are combined into a single energy tensor normally denoted by T_{ij}.

Coordinates

BEFORE DELVING into general relativity it is necessary to explain the possibility of more general forms of coordinates. In two dimensions we are familiar with rectangular coordinates x and y. The distance ds between two points whose x and y coordinates differ by dx and dy is of course given by

$$ds^2 = dx^2 + dy^2 \qquad \text{by Pythagoras}$$

and in three dimensions we have $ds^2 = dx^2 + dy^2 + dz^2$ in an obvious way.

Sometimes it is convenient to use other forms of coordinates such as polar coordinates where a point is described by r and θ where r is the distance of the point from some origin and θ is the angle from some fixed line. In this case the distance between two points whose coordinates differ by *small* values dr and $d\theta$ is

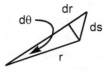

$$ds^2 = dr^2 + r^2 d\theta^2 \qquad \text{distance in polars}$$

The key point to notice is that the factor by which $d\theta^2$ is multiplied is not a constant but depends upon the value of the coordinates themselves (in this case just r).

Sometimes it is convenient to use oblique axes. Suppose the axes are at an angle ϕ as shown below. Then we can take the coordinates as the displacements parallel to the axes as in the diagram on the left. In this case the distance between two points is given by

$$ds^2 = dx^2 + dy^2 + 2 dx\, dy \cos \phi \qquad \text{displacements}$$

The interesting point here is that as well as the terms in dx^2 and dy^2 there is also a term in the cross-product $dx\, dy$. Such coordinates are often called contravariant coordinates.

An alternative view using oblique axes is to take the projections as the coordinates as in the diagram on the right. In this case the distance becomes

$$ds^2 = dx^2 / \sin^2\phi + dy^2 / \sin^2\phi - 2 dx\, dy \cos\phi / \sin^2\phi \qquad \text{projections}$$

Here the coefficients are quite different from those of the displacement model. These are often called covariant coordinates. Note that if the axes are rectangular then both forms of coordinates are the same.

It is time to use some adult notation rather than the x and y of the kindergarten. Suppose that there are n dimensions and that the coordinates are denoted by x with a suffix or superfix 1, 2, 3, etc. If we use the displacement (or contravariant) form then the coordinates are x^1, x^2, x^3, ..., whereas if we use the

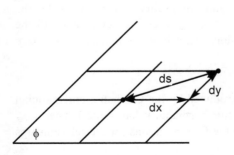

Displacement or contravariant coordinates.

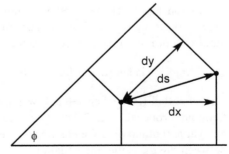

Projection or covariant coordinates.

projection (or covariant) form then the coordinates are denoted by $x_1, x_2, x_3, \ldots$. (Note that x^2 does not denote a power but simply the second coordinate in the displacement form.) The various coefficients of the coordinates can be denoted by g_{12} and g^{12} and so on. The two equations for the square of the interval ds^2 in the general two dimensional case can now be written

$$ds^2 = g_{11}dx^1dx^1 + g_{12}dx^1dx^2 + g_{21}dx^2dx^1 + g_{22}dx^2dx^2 \quad \text{displacement form}$$
$$ds^2 = g^{11}dx_1dx_1 + g^{12}dx_1dx_2 + g^{21}dx_2dx_1 + g^{22}dx_2dx_2 \quad \text{projection form}$$

We write dx^1dx^1 to denote the square of dx^1 to avoid ambiguity. Note carefully that the coefficients with the suffices g_{11} go with the coordinates with the superfices dx^1 and vice versa.

If we compare these equations with the two above we see that the values of the various gs for the oblique axes are

$g_{11} = 1;\ g_{12} = g_{21} = \cos \phi;\ g_{22} = 1$ \hspace{2em} displacements

$g^{11} = 1/\sin^2\phi;\ g^{12} = g^{21} = -\cos\phi/\sin^2\phi;\ g^{22} = 1/\sin^2\phi$ \hspace{1em} projections

It is convenient to make $g_{12} = g_{21}$ so that they are both half of the coefficient of the cross term $dx\ dy$.

It turns out that the upper and lower gs are always closely related no matter what coordinate system be chosen. We always have

$g_{11}g^{11} + g_{12}g^{21} = 1$ and $g_{11}g^{12} + g_{12}g^{22} = 0$
$g_{21}g^{11} + g_{22}g^{21} = 0$ and $g_{21}g^{12} + g_{22}g^{22} = 1$

Let us just test the first one for the oblique axes; we get

$g_{11}g^{11} + g_{12}g^{21} = 1 \times 1/\sin^2\phi + \cos\phi \times -\cos\phi/\sin^2\phi$
$= (1 - \cos^2\phi)/\sin^2\phi = \sin^2\phi/\sin^2\phi = 1$ \hspace{2em} OK

It is now time to introduce the summation convention. This states that if any expression has a repeated letter upstairs and downstairs then we consider the expression to be replaced by the sum of the terms where the letter takes all the values according to the number of dimensions. Thus in three dimensions we have

$a_{ij}b^{j2}$ \hspace{1em} is shorthand for \hspace{1em} $a_{i1}b^{12} + a_{i2}b^{22} + a_{i3}b^{32}$

The so-called Kronecker delta $\delta_i{}^j$ was introduced by the German mathematician Leopold Kronecker (1823-1891) as a useful shorthand. It has value 1 when $i = j$ and value 0 otherwise. We can now write the four equations for the relationships between the gs as the one equation

$g_{ij}\,g^{jk} = \delta_i{}^k$

and the two equations for the interval ds become

$$ds^2 = g_{ij} dx^i dx^j \quad \text{and} \quad ds^2 = g^{ij} dx_i dx_j$$

We see that the summation convention produces very compact notation – it is especially convenient when there are many dimensions.

The gs are very important for converting from the "upstairs" to the "downstairs" form and vice versa. For any vector a, it can be shown that

$$a_i = g_{ij} a^j \quad \text{and} \quad a^i = g^{ij} a_j$$

Thus we can convert between the dx^i and the dx_i using these rules.

The examples of coordinates we have used have been in the plane. Another interesting example is given by latitude and longitude which give coordinates on a sphere. Suppose the longitude is given by ϕ and the latitude by θ and that the radius of the sphere is a. Then the interval ds between two nearby points is given by

$$ds^2 = a^2 d\theta^2 + a^2 \cos^2\theta \, d\phi^2$$

from which the reader can easily find the gs by taking θ as x^1 and ϕ as x^2.

As another example, consider the equations of special relativity. Here we have three space dimensions and one time dimension and we can write

$$x^1 = x, \quad x^2 = y, \quad x^3 = z, \quad x^4 = t$$

The metric for special relativity

$$ds^2 = dx^2 + dy^2 + dz^2 - c^2 dt^2$$

can now be written as

$$ds^2 = g_{ij} x^i x^j$$

using the normal rectangular axes where $g_{11} = g_{22} = g_{33} = 1$ and $g_{44} = -c^2$ and the cross coefficients such as g_{12} are all zero.

Curvature

THE IMPORTANT TOPIC we are working towards is curvature. If we live on a sphere then we will find that our world is not flat. The angles of a triangle drawn on a sphere do not add up to 180° but are always more. (The sides have to be great circles which correspond to straight lines.) As an extreme case consider the triangle formed by taking the northern half of the 0° and 90° meridians and the section of equator joining them; all three angles of this triangle are right angles so that the sum of the angles is 270°. The surface of the sphere is intrinsically curved and it cannot be rolled out flat.

On the other hand the surface of a cylinder is quite different; it can be rolled out flat without any distortion. The surface of a cylinder is not intrinsically curved and the (two-dimensional) geometry of the surface is the same as that on a plane. A Flatlander living on a large cylinder would not know that it was not a plane from making measurements on the surface of the land – we assume that he doesn't travel too far and return to his starting point by going around his universe!

Of course we see the cylinder embedded in three-dimensional space and in that space it is curved. So there are different forms of curvature. When dealing with soap bubbles we were considering the curvature in the embedding space whereas here we are considering the intrinsic curvature that can be detected by measurements in the surface itself.

It is the gs that determine the properties of a space. An important feature is that the values of the gs can vary from place to place and in fact it is the ways the gs change that determine the curvature rather than the gs themselves. We thus need to introduce the rate of change of the g_{ij} with respect to the x^k. Normal calculus notation would be to write this as

$$\frac{\partial g_{ij}}{\partial x^k} \qquad \text{rate of change of } g_{ij} \text{ with respect to } x^k$$

This is rather cumbersome and is usually abbreviated as simply $g_{ij,k}$.

A number of mathematicians worked on the general form of n-dimensional space as an abstract concept in the 19th century. They included the German mathematician Riemann (1826–1866) and his pupils Christoffel and Ricci. One of the problems in defining curvature was finding a formula for the rate of change of a property with respect to the coordinates that was consistent in any set of coordinates. To cut a long story short the following rather complex expression was found to be important and is referred to as a Christoffel symbol

$$\Gamma^i_{jk} = \tfrac{1}{2} g^{il} (g_{lj,k} + g_{lk,j} - g_{jk,l}) \qquad \text{definition of Christoffel symbol}$$

It was discovered that the curvature can be described in terms of the so-called Riemann tensor defined as

$$R^i_{jkl} = \Gamma^i_{jl,k} - \Gamma^i_{jk,l} + \Gamma^i_{mk}\Gamma^m_{jl} - \Gamma^i_{ml}\Gamma^m_{jk} \qquad \text{definition of Riemann tensor}$$

It can be shown that if all the components R^i_{jkl} are zero then the space is flat and vice-versa. The Ricci tensor is a contracted form thus

$$R_{jk} = R^i_{jki} \qquad \text{definition of Ricci tensor}$$

Remember the rule that if a letter is repeated upstairs and downstairs then we have to sum over all possibilities. This means in four-dimensional space that

$$R_{jk} = R^1{}_{jk1} + R^2{}_{jk2} + R^3{}_{jk3} + R^4{}_{jk4}$$

The Ricci tensor has 10 independent components just as there are 10 g_{ij} for four-dimensional space. Given the g_{ij} we can compute the R_{ij} (just follow the definitions above) and similarly given the R_{ij} we can compute the g_{ij} (not so easy) and this gives us all we need to know about the space.

Incidentally, we can contract the Ricci tensor itself to give the scalar R. First we have to raise one of the suffices using the g^{ij} and then we can do the contraction thus

$R^i{}_j = g^{ik} R_{kj}$ raise one suffix of the Ricci tensor, then

$R = R^i{}_i = g^{ij} R_{ij}$ definition of R

R is a single value describing the overall curvature at a point.

Einstein's equations

EINSTEIN WAS NOT SATISFIED with special relativity because it did not explain gravitation. However, it did seem that light travelled in straight lines or perhaps we should say that the path of light defined straight lines. He thought that gravitation in some way affected the very fabric of space. After studying the geometry of Riemann, Christoffel and Ricci for many years, he proposed his general theory in 1915.

The basic equation for free space is simply

$R_{ij} = 0$ general relativity in free space

and the equation in the local presence of matter is

$R_{ij} - \frac{1}{2} g_{ij} R = k T_{ij}$ in the presence of matter, where k

$k = 8\pi G/c^4$ involves the gravitational constant G

The tensor T defines the quantity of energy/matter present. It includes both mass and electromagnetic radiation.

The more general equation reduces to the simple form if T_{ij} is zero. This is an interesting exercise in tensor manipulation. First we get

$R_{ij} - \frac{1}{2} g_{ij} R = 0$ then multiply by g^{ik} to give

$R_i{}^k - \frac{1}{2} g_{ij} g^{ik} R = 0$ then contract by putting $k = i$ to give

$R - 2R = 0$ note that $g_{ij} g^{ji} = \delta_i{}^i = 4$ since 4 dimensions

and so R is zero and can be omitted thereby leading to the simple equation.

The Schwarzschild solution

ALTHOUGH EINSTEIN'S EQUATIONS for general relativity in free space look remarkably simple in the compact form

$$R_{ij} = 0 \qquad \text{general relativity in free space}$$

they are really very complex. Exact solutions are hard to find but Schwarzschild (1873–1916) found an important solution in 1916 while serving on the Russian front. Using spherical coordinates r, θ, ϕ, where r is the distance from the origin and θ and ϕ are two angles corresponding to latitude and longitude, he showed that a solution was given by the metric

$$ds^2 = (1-2m/r)^{-1}dr^2 + r^2(d\theta^2 + \sin^2\theta\, d\phi^2) - (1-2m/r)\, dt^2 \quad \text{Schwarzchild}$$

It is instructive to compare Schwarzchild's metric with the normal formula for the Euclidean distance in spherical coordinates which is

$$ds^2 = dr^2 + r^2(d\theta^2 + \sin^2\theta\, d\phi^2) \qquad \text{Newton}$$

There are two key differences. One is that there is a time term but that is to be expected since we are dealing with relativity. But the really significant change is the factor $(1-2m/r)$ which modifies the terms in both the time, t, and the radial distance from the origin, r.

It turns out that the Schwarzschild metric describes the space around a central mass m where m is measured in suitable units. Using conventional units the term is

$$1 - 2Gm/rc^2 \qquad \text{distortion term in conventional units}$$

where G is the gravitational constant and c is the velocity of light. Remember that the gravitational constant G is such that the gravitational force between two bodies is

$$F = Gm_1m_2/r^2 \qquad \text{gravitational force between two bodies}$$

where m_1 and m_2 are the masses of the two bodies and r is the distance between them.

The interesting thing about the Schwarzschild metric is that when r has the value R defined by

$$R = 2Gm\,/\,c^2 \qquad \text{definition of Schwarzschild radius}$$

then the distorting factor becomes zero which means that the coefficient of dr in the metric becomes infinite and that of dt becomes zero predicting strange things.

In the case of the Sun, its mass is 2×10^{30} kg, the speed of light is 3×10^{8} metres/sec and the gravitational constant is 6.7×10^{-11}, so

$$R = 2\times6.7\times10^{-11}\times2\times10^{30} / (9\times10^{16}) = 3000 \text{ metres}$$

This is a very small distance compared with the radius of the sun which is nearly 700,000 km and so the distortion of space caused by the sun is almost negligible.

But what it does mean is that if all of the mass of the sun were compressed so that it fitted within this radius then odd things would happen. We now call such an object a Black Hole. But before discussing black holes we will briefly look at the other predictions of general relativity.

Consequences of general relativity

ONE FEATURE of the motion of the planets which could not be explained by Newton's laws concerned the orbit of Mercury. The orbits of the planets are basically ellipses with the Sun at one focus (strictly the focus is at the common centre of gravity but since the Sun is so massive compared with the planets it is inside the Sun anyway). The planets disturb each other's orbits a bit and as a consequence the ellipse shifts. We call this movement the motion of the perihelion – the perihelion is the point of the orbit nearest to the Sun.

The perihelion of Mercury shifts by almost exactly 56" per year. Remember that $60" = 1'$ and $60' = 1°$. Most of this shift can be explained by the effects of the other planets according to Newton's laws. But there is a discrepancy of 43" per *century* between observation and this prediction. This discrepancy was certainly known to Le Verrier who predicted the existence and location of Neptune by studying the discrepancies in the orbit of Uranus. Neptune was then discovered by the observatory in Berlin in 1846. But Le Verrier was quite unable to explain the discrepancy in the case of Mercury.

The distortion of space–time described by the Schwarzschild solution predicts that the orbit of a planet will rotate by

$3\pi R/a(1-e^2)$ movement of perihelion per orbit

In this formula, R is the Schwarzschild radius of the parent body (the Sun), a is the semi-major axis of the orbit (that is half the long axis of the ellipse) and e is the eccentricity of the ellipse. (If the equation of an ellipse is $x^2/a^2 + y^2/b^2 = 1$, then the eccentricity is given by $e^2 = 1 - b^2/a^2$.)

For the Sun, R is 3000 metres as mentioned above, the semi-major axis of the orbit of Mercury is 58,000,000 kilometres and the eccentricity of its orbit is 0.2. Putting these values in the equation we get

$$3\pi\times3000 / (58\times10^9 \times 0.96) = 0.508 \times10^{-6}$$

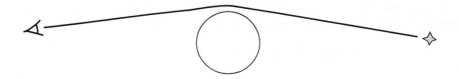

Light from a distant star is deflected when it passes close to the sun.

This is the rotation of the perihelion in radians per orbit. The period of Mercury is 88 days and there are 36524 days in a century so there are 36524/88 orbits per century which is about 415. To convert from radians to seconds of arc we first have to multiply by 57.3 ($180/\pi$) to convert to degrees and then by 3600 since there are 3600 seconds in a degree. We finally get

$0.508 \times 10^{-6} \times 415 \times 57.3 \times 3600 = 43$ seconds per century

and this exactly matches the observed discrepancy. This was the first verification of the theory of general relativity.

Another consequence is the bending of light by massive bodies. Of course light truly travels in "straight" lines in that it follows the path where $ds^2 = 0$. But it seems as if it is bent. This can be measured in an eclipse of the sun where the stars which can be seen just adjacent to the eclipsed sun appear shifted since the light from them is bent as it grazes the sun. The deflection predicted by general relativity is

$2R/r$ deflection of light grazing massive body

where R is the Schwarzschild radius and r is the actual radius of the Sun (about 695000 km). This gives $2 \times 3000/695000000 = 0.86 \times 10^{-5}$ radians $= 1.78"$.

There was a total eclipse of the Sun on 29th May 1919 and the observations made by expeditions to Sobral in Brazil and the Island of Principe which were organized by Eddington confirmed this prediction.

Another prediction is that clocks go more slowly in the presence of a gravitational field and this too has been confirmed.

Black Holes

THE TERM Black Hole seems to have been coined by John Wheeler in 1967. Prior to this time they were referred to as Collapsed Stars or Frozen Stars. Of course no-one really knew that they existed until very recently. Indeed, as recently as 1975, Dirac says in his book entitled *General Theory of Relativity* "The question arises whether such a region can actually exist. All we can say definitely is that the Einstein equations allow it."*

*Reproduced by permission from John Wiley & Sons Inc.

Einstein developed the theory of General Relativity in 1915 and, as mentioned above, Schwarzschild soon solved the equations for a sphere of mass *m*. The German mathematician Hermann Weyl (1885–1955) discusses this and other similar problems in a book published in 1921 so the mathematical possibilities have been in front of us for many years.

The main characteristics of a black hole is that, viewed from outside, time stands still at the surface and light cannot get out (photons stand still at the horizon). If we watch someone falling into a black hole then from our point of view they never quite get there; they seem to take an infinite time. For them, however, time goes all too quickly and they dash into the hole to an almost immediate death.

A black hole of the mass of the Sun has a radius of 3000 metres. In the case of the Earth, the corresponding radius is about one centimetre. So the whole of the Earth would have to be squeezed into a ball less than an inch across.

But a black hole doesn't have to be quite so amazingly dense. The relation between the mass m and the radius R is simply linear

$R = 2Gm / c^2$

So if we double the mass then the critical radius doubles as well. This means that a sufficiently large black hole doesn't have to be that dense. Suppose we had a huge incompressible sphere of the same density as water (1000 kg per cubic metre) and of radius R. Then its mass m is $4/3 \pi R^3 \times 1000$. Putting this in the equation for R we have

$R = 2G \times 4/3 \pi R^3 \times 1000 / c^2 = 6 \times 10^{-25} \times R^3$ from which

$R^2 = 1.5 \times 10^{24}$ giving

$R = 1.2 \times 10^{12}$ metres   or   1200 million kms

So a sphere of water of this size would form a black hole. This is the size of the Solar System nearly out to the orbit of Saturn so it's pretty huge but not incomprehensible. Of course a sphere of water this size would be crushed under gravity and be much denser. However, a group of ordinary stars rotating about a common centre can easily have an average density in order to form a black hole where the centrifugal force prevents them from collapsing. Rotating black holes seem to exist at the centre of many galaxies.

The other way to get a black hole is to have extremely dense matter and in fact a neutron star is an obvious example. (A neutron star is one composed just of neutrons crushed up together.) The mass of a neutron is about 1.7×10^{-24} gm, and the diameter of a neutron is about 10^{-13} cm. So the density is about 3×10^{15} times that of water. A black hole of this density using the same formula as before has a radius of 20 km compared to the Schwarzschild radius for the Sun of 3 km. So for a neutron star to become a black hole it has to have a mass about 7 times that of the Sun.

Despite Einstein's equations in 1915 showing the possibility, nobody really took black holes seriously for many years – Eddington dismissed the possibility as ridiculous. However, Oppenheimer (who worked on the wartime atomic bomb project) suggested around 1939 that neutron stars might exist and form what we now call black holes.

The discovery of Pulsars in 1968 which turned out to be rotating neutron stars was the first sign that black holes might be real. Now we know that black holes exist at the centre of many galaxies. Roger Penrose and Stephen Hawking have both done amazing pioneering work on the properties of black holes.

Properties of black holes

BLACK HOLES have some strange properties. The most obvious is that matter falling into a black hole cannot get out once it has passed inside the critical radius. This even applies to photons so that light cannot get out either.

The gravitational field bends light and this bending is such that light can circle for ever around a black hole in an orbit with a radius of $3R/2$ where R is the Schwarzschild radius of the black hole. Light a bit further out can be deflected by exactly 180° and so reflected back. So, in principle if we shine a torch on a black hole we will see a dark space surrounded by a halo of reflected light. This does not seem to have been observed as yet.

Some black holes rotate. This arises because stars rotate and a massive star which collapses into a black hole will preserve the angular momentum of the matter that is captured. The corresponding equations were solved by Roy Kerr and so rotating black holes are sometimes referred to as Kerr black holes.

It is usually stated that instant death is the fate of the traveller into a black hole because the massive variation in the gravitational field will tear him apart. But this is not necessarily so for huge black holes.

The Schwarzschild solution applies to the space outside an object, that is where there is no matter. However, it can be shown that in the case of matter distributed on the surface of a sphere, the effect outside is as if it were all concentrated at the centre and moreover, there is no gravitational field inside the sphere at all. This property of Newtonian physics perhaps surprisingly carries over into the relativistic solution as well.

So we can imagine a huge spherical shell of thickness sufficient to produce a black hole as observed from a distance yet within the shell the properties of space are as normal. Of course one could not get out of such a black hole but then maybe one wouldn't want to if for example it contained a small star with a comfortable planet rotating about it.

Such a shell-like distribution could perhaps be formed by many stars rotating about the centre of a galaxy. From within the centre of such an agglomeration things would seem much as normal, but from afar it would appear as a black hole at the centre of the galaxy.

Further reading

THERE ARE MANY popular accounts of relativity. One is *Black Holes* by Jean-Pierre Luminet. Despite the title it is not just about black holes. An excellent account of the basic principles with some simplified mathematical explanation is in *Makers of Mathematics* by Hollingdale; this book covers many other subjects discussed in these lectures and is highly recommended.

To go further into the mathematics is hard work. There is *The Meaning of Relativity* by Einstein himself. First published in 1922, the sixth edition is dated 1956. Another historic work is *Space Time Matter* by Hermann Weyl. The English translation is from the fourth German edition of 1921; this book contains a good account of tensors. A more recent work is *Essential Relativity* by Rindler. This has some light-hearted examples but also the real mathematics as well. The short book *General Relativity* by Dirac gives a brief but complete account of the necessary tensor notation and the essence of the key points of the theory; but it is more like printed college notes than a conventional book – lots of equations but not much chatter.

Finally, there is *Mr Tompkins in Wonderland* by George Gamow. This is the light-hearted account of the adventures of Mr Tompkins in a land where the speed of light is only about 30 mph. This will be found in *Mr Tompkins in Paperback* which also contains *Mr Tompkins Explores the Atom*.

Exercise

1 A very high speed train is 500 metres long as measured when stationary. The two power cars (one at each end of the train) have a device for making a mark on the track. In the centre of the train is a switch which sends a signal to the two ends of the train in an identical manner and causes the devices to make the marks on the track. One day the train is travelling at 0.866 times the speed of light; at this speed the Lorentz–Fitzgerald contraction factor is 0.5. The conductor throws the switch and the two marks are made on the track. An observer on the track measures the distance between the two marks. How far apart are they?

a) 500 metres

b) 1000 metres

c) 250 metres

d) 866 metres

10 Finale

THESE LECTURES have had a number of goals. One was simply to present some pretty or surprising configurations. Another was to reveal that despite the fact that we live in a three-dimensional world, nevertheless our understanding of three dimensions is fairly poor. Thus few people know that if you cut through a cube in a certain way, then the cross-section is a hexagon. Another goal, and perhaps the most important in a philosophical sense, was that solving and understanding a problem depends very much upon getting the right point of view.

The lecture on inversion and its use to prove Steiner's porism and explain Soddy's hexlet is perhaps the most intriguing example of getting the right point of view that we have encountered. Another example is that of the train marking the track in special relativity where it is important to analyse the situation from the point of view of the correct observer.

In this final lecture we will look at some more examples where getting the right point of view is so important. These are about certain curious properties of triangles, squares and other rectilinear figures and how the use of the Argand plane can provide very simple explanations in many cases.

Squares on a quadrilateral

IF WE TAKE any quadrilateral, place squares on its four sides, then the lines joining the centres of the opposite squares have the same length and are at right angles to each other.

Coxeter gives this as Exercise 10 in Section 1.8 of his *Introduction to Geometry*. It is preceded by two related problems and the three are as follows*:

8 If four squares are placed externally (or internally) on the four sides of any parallelogram, their centers are the vertices of another square.

9 Let X, Y, Z be the centers of squares placed externally on the sides BC, CA, AB of a triangle ABC. Then the segment AX is congruent and perpendicular to YZ (also BY to ZX and CZ to XY).

10 Let Z, X, U, V be the centers of squares placed externally on the sides AB, BC, CD, DA of any simple quadrangle ("quadrilateral") $ABCD$. Then the segment ZU (joining the centers of two "opposite" squares) is congruent and perpendicular to XV.

Coxeter gives brief answers (hints) which are reproduced below. In each case his answer is followed by further explanation which will be helpful (maybe essential) to mere mortals.

*Reproduced by permission from John Wiley & Sons Inc.

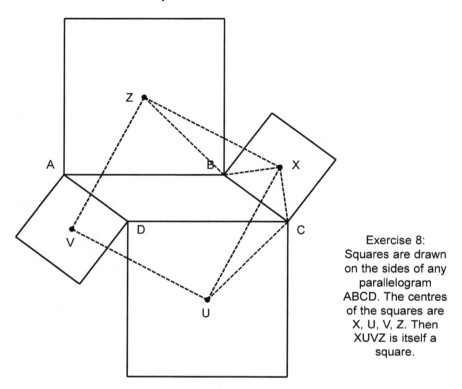

Exercise 8: Squares are drawn on the sides of any parallelogram ABCD. The centres of the squares are X, U, V, Z. Then XUVZ is itself a square.

Coxeter's answer for Exercise 8 shown above is:

8   Let Z, X, U be the centers of the squares on three consecutive sides AB, BC, CD of the parallelogram $ABCD$. The triangle XBZ is derived from XCU by a quarter-turn (i.e. a rotation through a right angle) about X.

The triangles XBZ and XCU are congruent because 1) $XC = XB$ both being half of the diagonals of the square centre X, 2) $BZ = CU$ both being half of the diagonals of the equal squares with centres Z and U, and 3) angle XBZ = angle XCU since both are two angles of 45° plus an acute angle of the parallelogram.

As a consequence since XB is at right angles to XC, the triangle is rotated a quarter-turn about X. And therefore XZ is at right angles to XU. Moreover, since the triangles are congruent XZ has the same length as XU.

Similarly, the other pairs of adjacent sides of the figure $ZXUV$ are at right angles and have the same length. Therefore the figure is a square.   QED

And now for Exercise 9 shown opposite. Here is Coxeter's answer:

9   Let M be the midpoint of CA. By Exercise 8, the segments MZ and MX are congruent and perpendicular. The same can obviously be said of MY and MA. Therefore the triangle MAX is derived from MYZ by a quarter-turn about M.

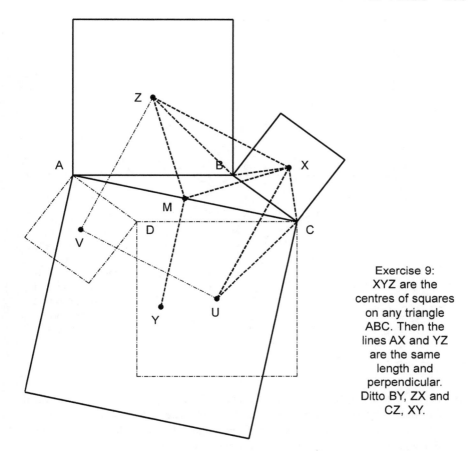

Exercise 9: XYZ are the centres of squares on any triangle ABC. Then the lines AX and YZ are the same length and perpendicular. Ditto BY, ZX and CZ, XY.

The first point to note is that looking back at the diagram for Exercise 8 we see that AC is a diagonal of the parallelogram. So M is the midpoint of the parallelogram and by symmetry (rotation) M is the midpoint of the square $ZXUV$. Therefore MZ and MX are half diagonals of the square and so of equal length and perpendicular.

MY and MA are both of the same length and are perpendicular because they are both equal to half the side of the square on AC.

The triangles MAX and MYZ are congruent because a) $MA = MY$, just proved, b) $MX = MZ$, just proved, and c) angle AMX = angle YMZ since they are both angle AMZ plus a right angle (ZMX and AMY respectively).

As a consequence since MA is at right angles to MY, the triangle is rotated a quarter-turn about M. And therefore AX is at right angles to YZ. Moreover, since the triangles are congruent AX has the same length as YZ. Similarly, BY is congruent to and perpendicular to ZX and CZ to XY. QED

And at last we are able to tackle Exercise 10 that the lines joining the centres of opposite squares on any quadrilateral are equal and perpendicular.

234 Gems of Geometry

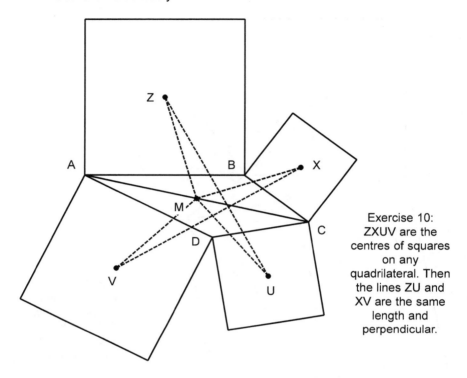

Exercise 10: ZXUV are the centres of squares on any quadrilateral. Then the lines ZU and XV are the same length and perpendicular.

10 As in Exercise 9, the segments *MZ* and *MX* are congruent and perpendicular. Similarly (by considering the triangle *CDA* instead of *ABC*), the segments *MU* and *MV* are congruent and perpendicular. Therefore the triangle *MXV* is derived from *MZU* by a quarter-turn about *M*.

The diagram is much as for Exercise 8 except that the point *D* has been moved so that *ABCD* is no longer a parallelogram but an arbitrary quadrilateral. The triangle *ABC* is as in Exercise 9. And so *MZ* and *MX* are congruent and perpendicular as proved in that exercise.

We now consider the triangle *ADC* instead of *ABC* and then by an identical argument we can prove that the lines *MU* and *MV* are congruent and perpendicular.

The triangles *MXV* and *MZU* are congruent because a) *MU* = *MV*, just proved, b) *MX* = *MZ*, just proved, and c) angle *VMX* = angle *UMZ* since they are both angle *UMX* plus a right angle (*UMV* and *ZMX* respectively).

As a consequence since *MU* is at right angles to *MV*, the triangle is rotated a quarter-turn about *M*. And therefore the third side *VX* is at right angles to *ZU*. Moreover, since the triangles are congruent *VX* has the same length as *ZU*. QED

Well, that was quite hard work and although we did not have to prove all of Exercise 9 on the way, nevertheless we did have to do the first part of it. We will now look at a stunning alternative approach.

The Argand plane

WE IMAGINE the diagram embedded in the complex plane or Argand diagram named after the Swiss mathematician Jean Robert Argand (1768–1822). Each point can be represented by the complex number at that point as a sort of coordinate. A complex number can be written as $x+iy$ where x and y are the so-called real part and imaginary part respectively and i is $\sqrt{-1}$; x and y then correspond to the normal coordinates in a plane.

So points such as A and B are represented by the complex numbers A and B. A line in the complex plane is characterized by the difference between the complex numbers representing its ends. So the line AB is represented by the value $B-A$. (It is important to get the sign right; considered in the opposite direction, the line BA is $A-B$.) Note that the point $B-A$ is the end of a parallel and equal line with one end at zero.

Note also that the midpoint of AB is $(A+B)/2$. A third of the way along is $(2A+B)/3$ and so on.

If a line is rotated anticlockwise by a right angle then its value is multiplied by i. So if two lines have values such that one is i times the other then they must have the same length and be at right angles.

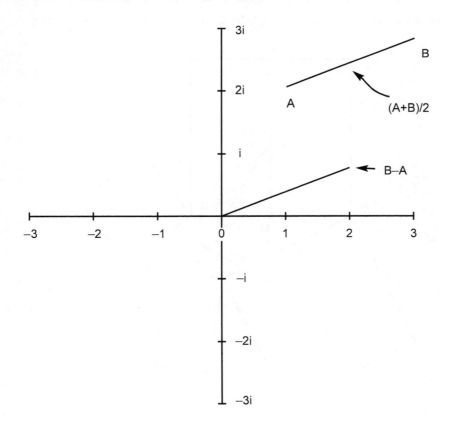

The quadrilateral revisited

THE QUADRILATERAL has points with values A, B, C, D. The midpoint of AB is M which thus is $(A+B)/2$.

The distance AM is $(B-A)/2$. Now MZ is the same length as AM and at right angles to it. So the line MZ is $(B-A)i/2$. The point Z is then M plus MZ, so

$$Z = (A+B)/2 + (B-A)i/2$$

Similarly

$$X = (B+C)/2 + (C-B)i/2$$
$$U = (C+D)/2 + (D-C)i/2$$
$$V = (D+A)/2 + (A-D)i/2$$

The line ZU is the difference between Z and U, and VX is the difference between V and X, so

$$ZU = (C+D)/2 + (D-C)i/2 - (A+B)/2 - (B-A)i/2 = (-A-B+C+D)/2 + (A-B-C+D)i/2$$
$$VX = (B+C)/2 + (C-B)i/2 - (D+A)/2 - (A-D)i/2 = (-A+B+C-D)/2 + (-A-B+C+D)i/2$$

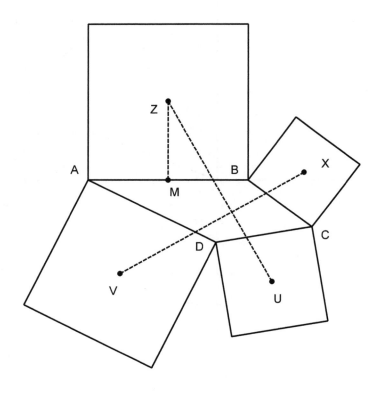

We now immediately see that $ZU \times i = VX$. It follows that ZU and VX are of the same length and at right angles. QED

Pretty smart stuff, eh? A good illustration of getting the right view of the problem or maybe one might say getting the right angle!

Other complex problems

A NUMBER of other curious problems can be solved using the Argand plane in much the same way. Some depend upon the observation that the centre of gravity (centroid) of a triangle ABC is the point $(A+B+C)/3$.

Consider any hexagon $ABCDEF$ as below. Draw the short diagonals between pairs of points thus AC, BD, CE, DF, EA and FB. This produces a number of triangles ABC, BCD, CDE, DEF, EFA, FAB. Take the centroids of these triangles (B', C', D', E', F', A') and join them to form another hexagon (in red). Then, whatever the shape of the original hexagon, this hexagon is such that its opposite sides are parallel and equal.

The proof is straightforward. The various points of the new hexagon are

$B' = (A+B+C)/3$

$C' = (B+C+D)/3$

and so on. It follows that

$B'C' = C'-B' = (B+C+D)/3 - (A+B+C)/3 = (D-A)/3$

$E'F' = F'-E' = (E+F+A)/3 - (D+E+F)/3 = (A-D)/3$

and the result immediately follows since the opposite sides are clearly of the same length and parallel.

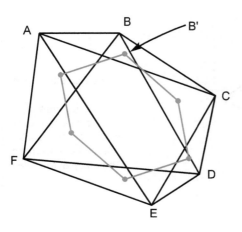

The opposite sides of the hexagon formed by the centroids of the triangles such as ABC are equal and parallel.

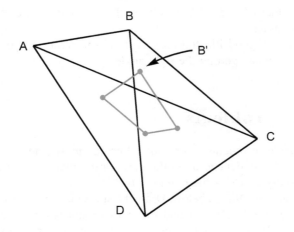

The quadrilateral formed by the centroids of the triangles such as ABC is similar to the original and one-third the size.

Take any quadrilateral $ABCD$ as above and join the diagonals in the same way so that, for example, B' is the centroid of the triangle ABC. Then the quadrilateral $A'B'C'D'$ is the same shape as the original but with sides of one-third the length.

Again the proof is straightforward. The side $B'C'$ is given by

$$B'C' = C'-B' = (B+C+D)/3 - (A+B+C)/3 = (D-A)/3$$

and so $B'C'$ is parallel to and one-third the length of the side DA of the original quadrilateral. The result follows.

Another rather trivial example is obtained by taking any triangle ABC and placing an equilateral triangle on each side as shown below. Then the centres of these equilateral triangles form a triangle $A'B'C'$ and the centroid of this triangle coincides with that of the original triangle.

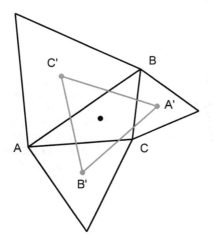

Napoleon's theorem.

The centres of equilateral triangles placed on any triangle themselves form an equilateral triangle.

The point C' is the centroid of the equilateral triangle on AB. This is obtained by taking the midpoint of AB and then going one-third of the way up the altitude which of course has length $\sqrt{3}/2$ times the side of the triangle and is at right angles to AB. We get

$$C' = (A+B)/2 + (B-A)i/2\sqrt{3}$$

If we add the three such centroids together, then the imaginary terms all cancel and we are left with $A+B+C$ and then dividing by 3 we get the centroid of $A'B'C'$ which is clearly the same as that of the original triangle ABC whose centroid is $(A+B+C)/3$.

Of more interest is that the triangle $A'B'C'$ is always equilateral no matter what the shape of ABC. This is known as Napoleon's theorem after Napoleon Bonaparte (1769–1821) who was an amateur mathematician as well as a general although it is unclear whether he really discovered it.

In order to prove this we need to consider the cube roots of 1 which we discussed when dealing with Newton's method in the lecture on Chaos and Fractals. We saw that the three cube roots of 1 are 1 and $-1/2 \pm \sqrt{3}i/2$. The value $-1/2 + \sqrt{3}i/2$ is often known as ω so that the three roots are 1, ω and ω^2. If two lines are at 60° to each other and of the same length then it is clear that the value of one will be ω times the value of the other (strictly $-\omega$ or $-\omega^2$).

The line $C'A'$ is given by

$$C'A' = A' - C' = (B+C)/2 + (C-B)i/2\sqrt{3} - (A+B)/2 - (B-A)i/2\sqrt{3}$$
$$= (C-A)/2 + (C+A-2B)i/2\sqrt{3}$$

If we multiply this by ω we get

$$\omega C'A' = ((C-A)/2 + (C+A-2B)i/2\sqrt{3}) \times (-1/2 + \sqrt{3}i/2)$$
$$= (A-C)/4 - (C+A-2B)/4 + i \times ((C-A)\sqrt{3}/4 - (C+A-2B)/4\sqrt{3})$$
$$= (B-C)/2 + (B+C-2A)i/2\sqrt{3} = B'C'$$

and so it follows that the line $C'A'$ has the same length as $B'C'$ and is at 60° to it. Therefore the triangle $A'B'C'$ is equilateral.

Napoleon's theorem is in fact a special case of a more general theorem which states that if we place *similar* triangles on the edges of any triangle then the circumcentres of these three triangles form a triangle similar to the three triangles. A traditional proof of this theorem is given in *Geometry Revisited* by Coxeter and Greitzer. Napoleon's theorem is of course the special case where the similar triangles are in fact equilateral in which case the circumcentre (the centre of the circle through the three points of the triangle) becomes just the centre or centroid.

Trisection

IF WE TRISECT the three angles of any triangle then the trisecting lines meet in points which form an equilateral triangle. This is known as Morley's theorem and was discovered as recently as 1899 by Frank Morley (1860–1937).

The odd thing about this theorem is that there appears to be no straightforward proof. All proofs seem to sort of work backwards from an equilateral triangle and show that any shaped triangle can be fitted around. Such a backward proof will be found in *Introduction to Geometry* by Coxeter. The reader is invited to become famous by devising a forward proof!

It is perhaps surprising that such a simple theorem was not discovered earlier. Maybe it was because it was known that trisection of an angle could not be done using ruler and compasses and therefore it seemed immoral to even think about it.

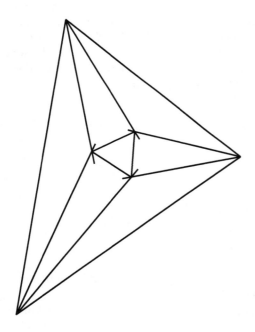

Morley's theorem.

The triangle formed by the trisectors of the angles of any triangle is equilateral.

Bends

IT IS OFTEN HELPFUL to talk about the bend of a circle rather than its radius. The bend is the inverse of the radius. The smaller the circle the bigger the bend which makes sense in that the circumference of a smaller circle is more bent. So the rule about the intermediate film where two soap bubbles meet which is normally written as

$1/R_{12} = 1/R_1 - 1/R_2$ rule for soap bubbles in terms of radii

can instead be written as

$B_{12} = B_1 - B_2$ rule for soap bubbles in terms of bends

Using bends is also neater for the sizes of the spheres in Soddy's hexlet since in many positions the bends are simply integers. Moreover, the sum of the bends of the three fixed spheres always equals the mean of the bends of any opposite pair of the hexlet. So the sum of the bends of the whole chain stays the same as the chain moves.

If four circles touch each other as shown below (like a Steiner chain of three), then the four bends have an interesting property, namely

$$2(b_1^2 + b_2^2 + b_3^2 + b_4^2) = (b_1 + b_2 + b_3 + b_4)^2$$

According to Coxeter, this was first discovered by the French mathematician René Descartes (1596–1650). He mentioned it in a letter to Princess Elisabeth of Bohemia in 1643. It was rediscovered in 1842 by Philip Beecroft and then again by Soddy in 1936 who wrote a poem about it in *Nature**. The middle verse is

> Four circles to the kissing come,
> The smaller are the benter.
> The bend is just the inverse of
> The distance to the centre.
> Though their intrigue left Euclid dumb
> There's now no need for rule of thumb.
> Since zero bend's a dead straight line.
> And concave bends have minus sign,
> *The sum of the squares of all four bends*
> *Is half the square of their sum.*

The reference to the negative bends reflects the fact that one circle might surround the others in which case that circle has a negative bend.

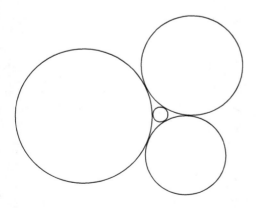

Four circles touching each other.

*Reprinted by permission from Macmillan Publishers Ltd: Nature, vol 137, p 1021, 1936.

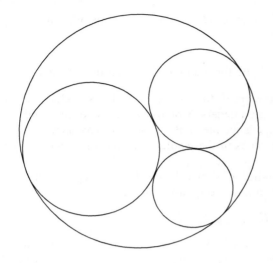

Four circles touching with one external.

Each set of four touching circles has a companion set through the points of contact as shown below in red. If the bends of one set are a_1, a_2, a_3, and a_4 then it turns out that these bends are related to the bends b_1, b_2, b_3, and b_4 of the other set by relations such as

$$2a_1 = b_2 + b_3 + b_4 - b_1 \qquad \text{etc}$$

and moreover, the sum of the bends of one set is equal to the sum of the bends of the other set, thus

$$a_1 + a_2 + a_3 + a_4 = b_1 + b_2 + b_3 + b_4$$

Soddy discovered that there is an analogous theorem and formula for five spheres in three dimensions and this is mentioned in the next verse of the poem. It seems that Thorold Gosset (1869–1962) extended it to $n+2$ hyperspheres in n dimensions and added a final verse.

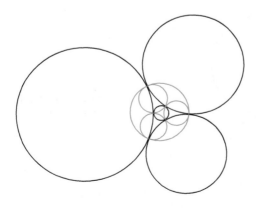

Four circles touching and their companion set.

Pedal triangles

HERE IS A LITTLE PUZZLE for the reader to struggle with. Consider any triangle ABC and a point P. It is usual to consider the case where P is inside the triangle. Drop the perpendiculars from the point P on to the three sides of the triangle. The feet of the perpendiculars form a triangle $A_1B_1C_1$ which is known as the pedal triangle with respect to the point P. (A special case is where the perpendiculars go through the vertices of the original triangle in which case P is the orthocentre.)

We find in general that the angles of the pedal triangle are different from those of the original triangle. Now starting from the pedal triangle find its pedal triangle with respect to the same point P. We might call this the grandpedal triangle of the original triangle. Finally, take the pedal triangle of this triangle which we can call the greatgrandpedal triangle of the original triangle.

The problem is to show that the angles of the greatgrandpedal triangle $A_3B_3C_3$ are the same as those of the original triangle ABC. In other words the 3rd generation pedal triangle is similar to the original triangle.

A similar construction can be done with a quadrilateral. Drop the perpendiculars from any point P to the four sides and then the feet of these define the vertices of another quadrilateral. The 4th generation such quadrilateral is similar to the original. In fact the theorem extends to a general polygon. The nth generation pedal n-gon of any n-gon is similar to the original.

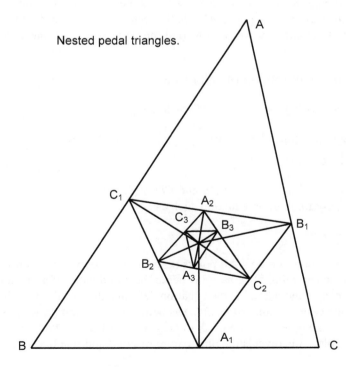

Nested pedal triangles.

Coordinates of points and lines

WHEN DEALING with projective geometry we saw that a point in a plane could be denoted by coordinates of the form (p, q, r) and that it was only the ratios that mattered. Thus (p, q, r) is the same point as $(2p, 2q, 2r)$. There is of course no point $(0, 0, 0)$. We can choose any three points not on a line as the triangle of reference $(1, 0, 0)$, $(0, 1, 0)$, $(0, 0, 1)$ and then any point not on the lines of the triangle as the unit point $(1, 1, 1)$.

We can use these coordinates both for "normal" geometry and for finite geometry. In the 13-point finite plane with arithmetic modulo 3, it means that the point $(1, 1, 2)$ can equally be denoted by $(2, 2, 4)$ which is $(2, 2, 1)$. Similarly, the point $(1, 2, 0)$ is the same as $(2, 1, 0)$.

If A and B are two points then any point on the line joining them can be written as $pA + qB$. It is only the ratio of p and q that matters.

If we need to find the point where two lines AB and CD meet where A, B, C, D are given points then there are a number of ways we can proceed. One is to observe that any point on AB can be written as $A + pB$ and any point on CD can be written as $C + qD$. If these are the same point then we know that the ratios of the coordinates must be the same. So we must have

$$A + pB = \alpha(C + qD) \qquad \text{equations for } p \text{ and } q \text{ and factor } \alpha$$

then by comparing the three coordinates we obtain equations for p and q.

As a simple example suppose ABC is the triangle of reference and D is the unit point $(1, 1, 1)$ and that we wish to find the point where AB meets CD. We obtain

$$(1, 0, 0) + p(0, 1, 0) = \alpha(0, 0, 1) + \alpha q(1, 1, 1)$$

and then by comparing the three coordinates, we get

$$1 = \alpha q; \quad p = \alpha q; \quad 0 = \alpha + \alpha q \qquad \text{from which}$$
$$q = -1; \quad \alpha = -1; \quad p = 1$$

so that the point of intersection of AB and CD is $(1, 1, 0)$.

We normally think of an equation such as

$$lx + my + nz$$

as being the equation of a line. All points (x, y, z) that satisfy the equation lie on the line. But we can think of (l, m, n) as being the coordinates of a line and then the equation can be thought of as the equation defining all lines through the point (x, y, z). In one case we think of (l, m, n) as being constant and (x, y, z) as being variable and in the other case we think of (x, y, z) as being constant and (l, m, n) as being variable. This reveals the dual nature of point and line once more.

In the Fano plane the seven points are 100, 010, 001, 011, 101, 110, 111. Equally, there are seven lines and these have the same set of seven coordinates. The line through the points 100, 101, 001 is the line 010. Thus the points of the triangle of reference are the points 100, 010, 001 and the lines of the triangle of reference are the lines 100, 010, 001. The unit point 111 is the only point not on a line of the triangle of reference and the unit line 111 (the "circle") is the only line not through a point of the triangle of reference.

A neat way to deal with coordinates is to use determinants. Remember that a determinant is a square array of numbers and has a value obtained by adding or subtracting products of the numbers taken one from each row and column. Thus in the simple 2×2 case

$$\begin{vmatrix} a & b \\ c & d \end{vmatrix} \qquad \text{2 by 2 determinant}$$

the value is defined to be $a \times d - b \times c$.

In the 3×3 case we might have

$$\begin{vmatrix} a_{11} & a_{12} & a_{13} \\ a_{21} & a_{22} & a_{23} \\ a_{31} & a_{32} & a_{33} \end{vmatrix} \qquad \text{3 by 3 determinant}$$

and the value is

$$a_{11}(a_{22}a_{33} - a_{23}a_{32}) - a_{12}(a_{21}a_{33} - a_{23}a_{31}) + a_{13}(a_{21}a_{32} - a_{22}a_{31})$$

which can be expressed in many different ways. Note how each term of the first row is multiplied by the 2×2 determinant obtained by omitting the row and column containing the term in the first row. The signs alternate.

Now if we are using x, y, z as coordinates then the equation of the line joining (x_1, y_1, z_1) to (x_2, y_2, z_2) is

$$\begin{vmatrix} x & y & z \\ x_1 & y_1 & z_1 \\ x_2 & y_2 & z_2 \end{vmatrix} = 0$$

So the coordinates of the line are the 2×2 determinants

$$\begin{vmatrix} y_1 & z_1 \\ y_2 & z_2 \end{vmatrix} \qquad \text{etc.}$$

The condition for three points to lie on a line is simply

$$\begin{vmatrix} x_1 & y_1 & z_1 \\ x_2 & y_2 & z_2 \\ x_3 & y_3 & z_3 \end{vmatrix} = 0$$

Similar relationships hold for line coordinates and the condition for three lines to go through a point.

It all works for the finite geometries as well provided we remember to do the appropriate modulo arithmetic.

Further reading

THE TRADITIONAL PROOF of the theorem about squares on a quadrilateral will be found in *Introduction to Geometry* by Coxeter; this book also describes the bends and touching circles. Poems by Soddy and Gosset about the bends will be found in *The Mathematical Magpie* compiled by Clifton Fadiman. Proofs of Napoleon's theorem, pedal triangles and the trisection theorem will all be found in *Geometry Revisited* by Coxeter and Greitzer. The property of hexagons is mentioned in the article *New Names for Old* by Kasner and Newman in volume 3 of *The World of Mathematics* by James Newman.

The use of determinants for projective geometry is described in *The Methods of Plane Projective Geometry based on the use of General Homogeneous Coordinates* by E A Maxwell.

A The Bull and the Man

WE HAVE SEEN that the square root of 2 and the golden number τ occur frequently in the geometry of various figures. Thus a diagonal of a square of unit side has length √2 whereas a diagonal of a pentagon of unit side has length τ. Generally, √2 crops up in the series of figures related to the cube and octahedron whereas τ crops up in the series of figures related to the dodecahedron and icosahedron. We will now look at a very different problem where both numbers occur.

The problem

SUPPOSE we have a square field with a gate in one corner as shown below. The field also contains a man and a bull. The bull is in a corner diagonally opposite the gate whereas the man is in one of the other corners and thus equidistant from the bull and the gate.

The man runs towards the gate. If the bull is smart then it will run diagonally across the field to the gate. The distance the bull has to go is obviously √2 times the distance the man has to go. So if the man can run at 10 miles per hour then the bull has to do about 14.14 miles per hour to catch him which sounds too close for comfort.

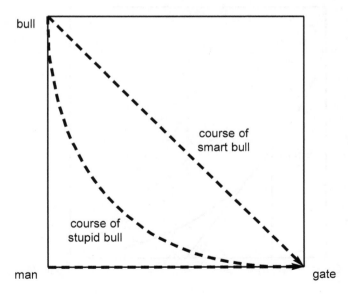

J. Barnes, *Gems of Geometry*, DOI 10.1007/978-3-642-30964-9,
© Springer-Verlag Berlin Heidelberg 2012

248 Gems of Geometry

However, if the bull is stupid then it will run towards the man and pursue a curved course across the field as the man runs along the edge of the field. If the bull gets to the gate at the same time as the man then the distance the bull travels is surprisingly τ times the length of the side of the field. So the bull has to travel at about 16.18 miles per hour. Hmm, still sounds dodgy. The proof is straightforward but somewhat tedious.

The proof

SUPPOSE the field is the unit square. The bull starts at $A\,(0, 1)$. The man starts at the origin $(0, 0)$. The gate is at $(1, 0)$.

Suppose the bull travels at speed λ times the man. The velocity of the bull is always directly towards the man. We use the normal x, y coordinates and let s be the distance the bull has travelled at any time.

Then the curve the bull traverses can be parameterized as

$$x = x(s), \quad y = y(s)$$

and we have the usual equations

$$ds^2 = dx^2 + dy^2$$

or

$$(ds/dy)^2 = (dx/dy)^2 + 1$$

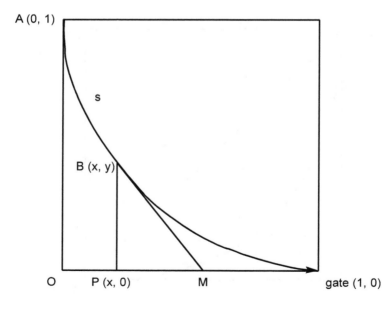

A The Bull and the Man

Suppose the bull is at the point $B\ (x, y)$. Then the man is at the point M where the tangent to the path of the bull meets the x-axis. So the distance PM is

$$-y\frac{dx}{dy} \quad \text{negative sign because slope is negative}$$

So the distance OM which the man has travelled is $x - y\,dx/dy$. But the distance s along the curve AB which the bull has travelled is λ times this because the bull is λ times faster than the man. So

$$\lambda\ (x - y\,dx/dy) = s$$

Now differentiate with respect to y

$$\lambda\ (dx/dy - y\,d^2x/dy^2 - dx/dy) = ds/dy = -\sqrt{(1 + (dx/dy)^2)}$$

Note that we take the negative square root since ds/dy is negative. So

$$\lambda\ y\,d^2x/dy^2 = \sqrt{(1 + (dx/dy)^2)}$$

Now put $p = dx/dy$. We get

$$\lambda\ y\,dp/dy = \sqrt{(1 + p^2)}$$

Rearrange and integrate

$$\lambda \int \frac{dp}{\sqrt{(1 + p^2)}} = \int \frac{dy}{y}$$

This is all standard stuff, we get

$$\lambda \sinh^{-1} p = \log y + k$$

where k is the constant of integration. Now we know that at the start, $y = 1$ and moreover, the bull goes straight down the y-axis so $p = dx/dy = 0$. So

$$\lambda \sinh^{-1} 0 = \log 1 + k$$

Now $\sinh^{-1} 0 = 0$ and $\log 1 = 0$ so $k = 0$ as well. Hence we get

$$p = \sinh((\log y)/\lambda)$$

Now replace p by dx/dy, also put $\mu = 1/\lambda$ to make life easier

$$dx/dy = \sinh(\mu \log y)$$

Integrate once more

$$x = \int \sinh(\mu \log y)\,dy$$

Now use

$$\sinh z = \tfrac{1}{2}(e^z - e^{-z})$$

this gives

$$x = \tfrac{1}{2}\int (e^{\mu \log y} - e^{-\mu \log y})\,dy$$

But of course $e^{\mu \log y}$ is simply y^μ. So

$$x = \tfrac{1}{2}\int (y^\mu - y^{-\mu})\,dy$$

This is a trivial integration and we get

$$2x = y^{1+\mu}/(1+\mu) - y^{1-\mu}/(1-\mu) + k$$

where again k is the constant of integration.

But we know two points on the curve, where the bull starts and the gate. So when $x = 0$, $y = 1$ and when $x = 1$, $y = 0$. Put these pairs of values in the above and we get two equations for μ and k.

$$0 = 1/(1+\mu) - 1/(1-\mu) + k$$
$$2 = k$$

Substituting $k = 2$ in the first equation and rearranging

$$2(1+\mu)(1-\mu) = (1+\mu) - (1-\mu)$$

which reduces to

$$1 - \mu^2 = \mu$$

Putting $\mu = 1/\lambda$, this becomes

$$\lambda^2 - 1 = \lambda$$

So finally

$$\lambda = \tau, \text{ the golden number.} \hspace{4em} \text{QED}$$

To finish we will put the values of $k = 2$ and $\mu = 1/\tau$ into the equation for x. We get, after noting that $1 + 1/\tau = \tau$ and $1 - 1/\tau = 1/\tau^2$,

$$x = \tfrac{1}{2}(y^\tau/\tau - \tau^2 y^{1/\tau^2}) + 1$$

In France this curve is known as the *courbe du chien*. Interesting variations arise if the angry bull (or mad dog) starts from other places such as (1, 1) or the man attempts other strategies such as running in a circle.

B Stereo Images

THIS APPENDIX presents stereo images of several of the compound figures described in Lecture 2 and the Desargues configuration discussed in Lecture 4. Other stereo images relating to projections of objects from four dimensions into three will be found in Appendix D.

These images have been designed to be viewed from about 30 cm. The left eye should look at the left image and the right eye at the right image. Some find it helpful to hold a strip of card in the middle so that each eye can only see the correct image.

The trick is to persuade your eyes that you are looking at something more or less at infinity so far as the muscles which align the eyes are concerned but to focus on the images which are quite near. If you are short-sighted it could be better to remove your spectacles and hold the pages somewhat closer. If you are normal or long-sighted then reading glasses will be found helpful.

Compound figures

In Lecture 2 we described a number of compound figures such as the Stella Octangula and the amazing compound of five cubes.

We start with simple images of cubes and tetrahedra.

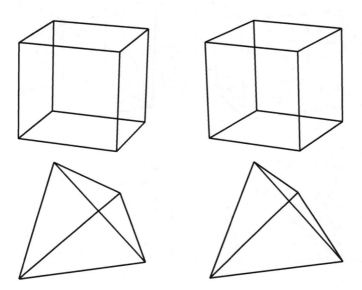

A tetrahedron can be inscribed in a cube in two different ways as in the first two images below. If both tetrahedra are inscribed in a cube and then the cube is removed we get the compound of two tetrahedra or the stella octangula as shown in the third and fourth images.

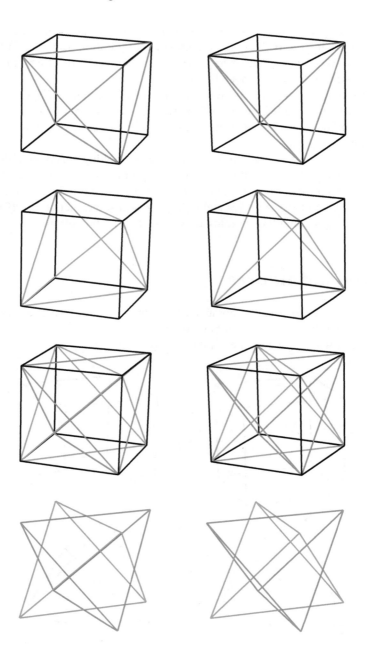

The space common to the two tetrahedra forming a stella octangula is in fact an octahedron. This is shown in thinner lines in blue in the first image below.

Deleting the hidden edges makes the structure of the stella octangula much clearer. The edges of the octahedron form the concave edges of the stella octangula.

Finally, we see how the octahedron is circumscribed by the original cube. The vertices of the octahedron lie in the centres of the faces of the cube.

Similarly, a cube can be inscribed in an octahedron. The vertices of the cube lie in the centre of the faces of the octahedron.

And of course the nesting can be continued indefinitely. In the bottom figure the inner cube is one-third of the size of the outer cube.

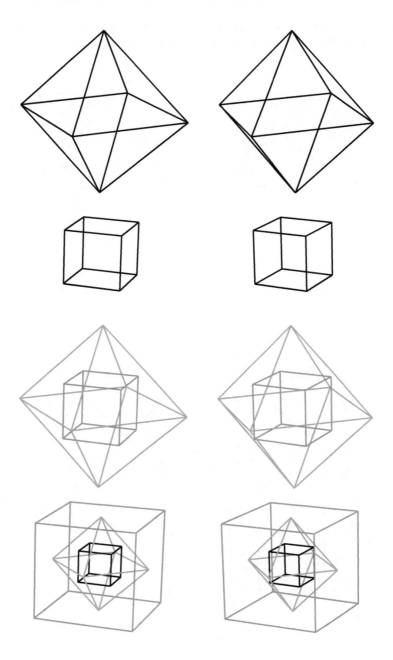

These images illustrate the compound of the cube and octahedron. The common space is a cuboctahedron. This is shown in thinner lines in blue.

The bottom image shows the compound with hidden lines removed. The edges of the cuboctahedron form the concave edges of the compound.

Similar images of the compound of icosahedron and dodecahedron can be drawn showing the common icosidodecahedron.

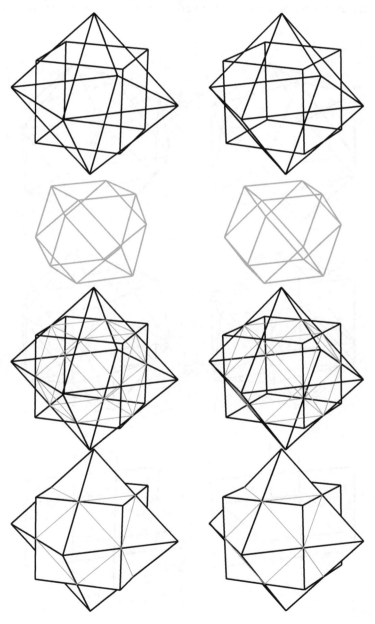

256   Gems of Geometry

The three images below show the icosahedron and the three inscribed golden rectangles. The top images opposite show a dodecahedron with an inscribed cube. The lower images show the three double golden rectangles inscribed in a dodecahedron.

The next two pages show the five positions of the cube in a dodecahedron and the building of the compound of five cubes as each cube is added in turn. The colours are as in Lecture 2 except that yellow is replaced by brown and white by black.

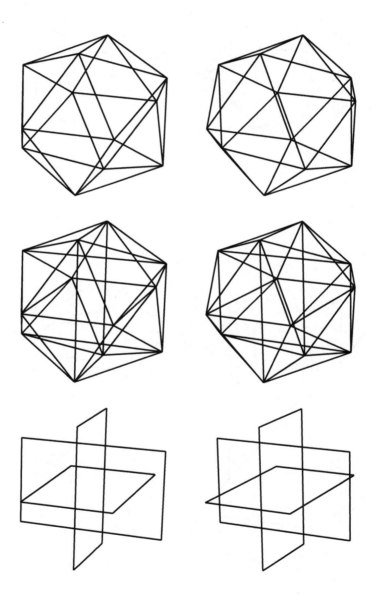

B Stereo Images

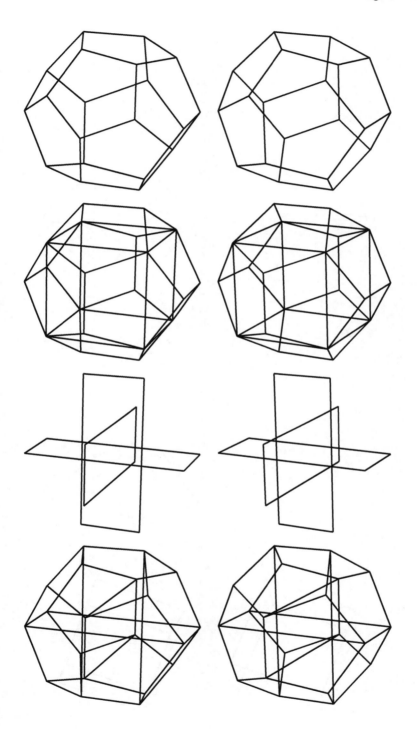

258 Gems of Geometry

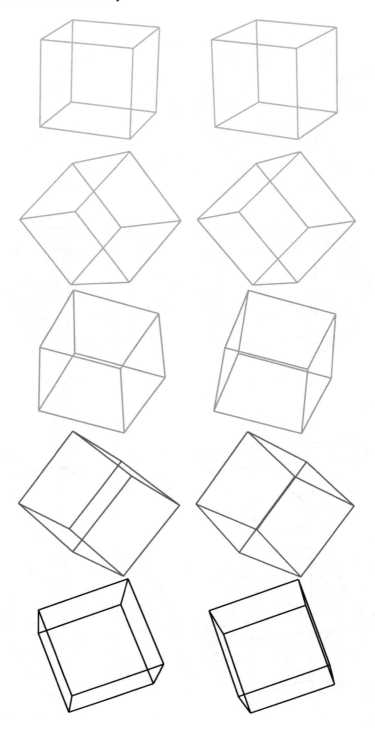

B Stereo Images 259

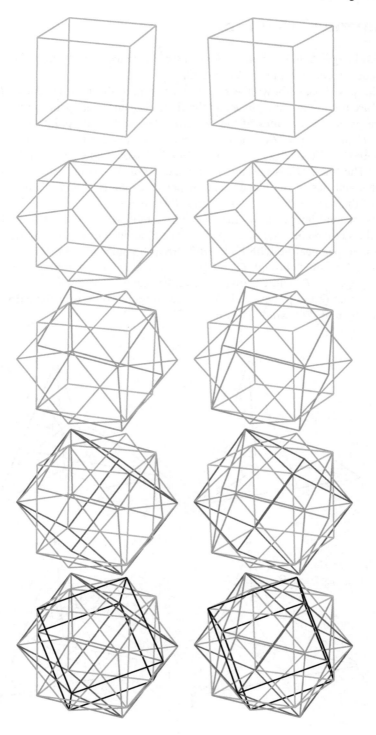

Desargues' theorem

The two images below are stereo versions of those used to illustrate the proof of Desargues' theorem on pages 91 and 92.

The first image shows the general case where the two triangles ABC and $A'B'C'$ are not in the same plane and the figure is a complete pentahedron formed by five planes. The planes of the two triangles contain six points each and are $ABCLMN$ and $A'B'C'LMN$. The three other planes involve the point of projection P and are $PBB'CC'L$, $PCC'AA'M$ and $PAA'BB'N$. Note that each point lies in three planes. The line LMN is the line of intersection of the planes containing the two triangles and lies in the plane of the paper. The points C and C' are above the plane of the paper whereas A, A', B and B' are below.

The second image shows the case where the two triangles ABC and $A'B'C'$ are in the same plane (that of the paper). The blue lines form the triangle $A''B''C''$ which is in perspective with both ABC (from the point Q) and $A'B'C'$ (from the point Q'). The lines through Q and Q' are in red. The red and blue lines and the points Q, Q', A'', B'' and C'' are all above the plane of the paper.

Note that Q and Q' have to lie on a line through P otherwise the pairs of lines such as AQ and $A'Q'$ would not meet.

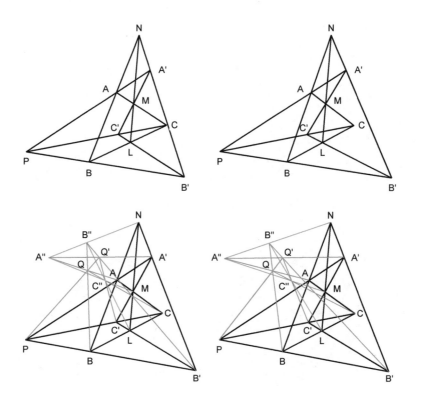

C More on Four

THIS APPENDIX looks at a few further aspects of regular and semi-regular figures in four dimensions.

Archimedean figures in four dimensions

I HAD SORT OF PROMISED not to look at this topic but a brief discussion might be interesting. We can define Archimedean figures in four dimensions as ones in which each cell is a regular Platonic solid but not all cells are the same. This is the obvious analogue of the rule in three dimensions which requires each face to be a regular polygon but not all faces to be the same.

We also need rules regarding the arrangement of cells around vertices, edges and faces. Recall that in three dimensions the rule is that the combination of faces around each vertex shall be the same and arranged in the same way. This gives rise to the 13 figures described in Lecture 2. We noted that if in addition we required the faces around each edge to be the same then only two figures qualified, namely the cuboctahedron and icosidodecahedron.

In four dimensions we obviously require that the arrangement of cells around each vertex has to be the same. We might find that some special figures have the same arrangement around each edge and face as well.

In Lecture 3 we introduced the 24-cell by considering the effect of taking a cubic honeycomb in various dimensions and subdividing alternate cubes and adding the portions to the surrounding cells. In two dimensions this just gave another tiling of squares; in three dimensions it gave the honeycomb of rhombic dodecahedra; in four dimensions it gave the honeycomb of 24-cells.

We can do a similar trick with other configurations. Take for example a triangle and consider the figure obtained by joining the midpoints of each side. We just get another triangle as shown below which is rather boring.

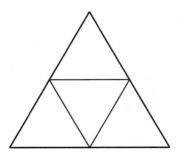

Joining the midpoints of the sides of a triangle just gives another triangle.

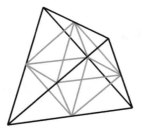 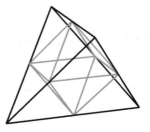

Joining the midpoints of the edges of a tetrahedron results in an octahedron.

Now consider what happens in three dimensions if we join the midpoints of the edges of a tetrahedron. We must take care only to join adjacent points, that is those points nearest to each other. If the edge of the tetrahedron has length 2 then the adjacent midpoints are 1 apart. We easily see that the resulting figure is an octahedron as shown above in stereo. Four faces lie in the four faces of the tetrahedron and four faces lie parallel to them.

We will now do the analogous construction in four dimensions. The analogue of a tetrahedron is of course the 4-simplex or 5-cell. This has 10 edges and so the figure obtained by joining the midpoints of these edges has 10 vertices. None of the four-dimensional figures we have encountered so far has 10 vertices so this is obviously something new.

We start by considering a stereo image of a 4-simplex as shown below. This is obtained by projecting the four-dimensional figure onto three dimensions. The result is a tetrahedron with a fifth point in the centre. Such projections are considered in more detail in Appendix D.

If we now join the midpoints of the edges the result is the rather confusing image shown opposite above. However, with care we see that the figure has 10 cells. Five cells are tetrahedra and five cells are octahedra. Three octahedra and two tetrahedra meet at each vertex.

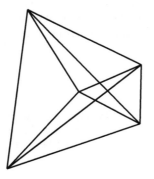 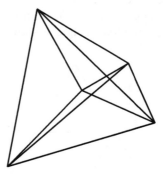

A stereo image of a 4-simplex.

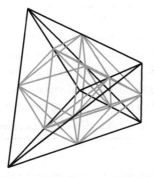 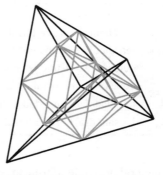

The result of joining the midpoints of the edges of the 4-simplex.

This becomes clearer if we remove the original 4-simplex and colour one octahedron and one tetrahedron as shown below. The four other tetrahedra have as base one face of the red octahedron and as apex one vertex of the blue tetrahedron. The four other octahedra lie between the other faces of the red octahedron and the faces of the blue tetrahedron. Note that the four tetrahedra look inverted; this is because the projection is orthogonal. If we had projected from much nearer then the blue tetrahedron would have been much smaller.

So this is a four-dimensional Archimedean figure. The 30 faces are all triangles but are not all equivalent. Some faces separate an octahedron and a tetrahedron whereas others separate two octahedra. There are also 30 edges with one tetrahedron and two octahedra around each edge. So this figure has a pleasing duality with 10 cells and vertices and 30 faces and edges.

The figure is somewhat analogous to the truncated tetrahedron which has four triangles and four hexagons as faces and in which each triangle is opposite a hexagon. In the four-dimensional figure each tetrahedron is similarly opposite an octahedron.

The figure can also be constructed by considering the truncation of a 4-simplex. Since the vertex figure of a 4-simplex is a tetrahedron it follows that the figures obtained at the truncation points are tetrahedra. The cells which also happen to be tetrahedra are first truncated into truncated tetrahedra and finally into octahedra.

The red octahedron is opposite the blue tetrahedron.

Object	Intermediate	Final truncation
4-simplex	tetrahedra + trunc tetrahedra	5 tetrahedra + 5 octahedra
Hypercube	tetrahedra + trunc cube	16 tetrahedra + 8 cuboctahedra
16-cell	octahedra + trunc tetrahedra	8 octahedra + 16 octahedra
24-cell	cubes + trunc octahedra	24 cubes + 24 cuboctahedra
120-cell	tetrahedra + trunc dodecahedra	600 tetrahedra + 120 icosidodecahedra
600-cell	icosahedra + trunc tetrahedra	120 icosahedra + 600 octahedra

The general principle of truncation can be applied to the other regular figures in the same way. The new cell which emerges at a truncated vertex is of course the vertex figure while the effect of truncation on the original cells is exactly as happens with truncation of those cells in three dimensions.

Thus consider the hypercube. The vertex figure of a hypercube is a tetrahedron and so the new cells created at the 16 vertices are tetrahedra. The 8 cells of a hypercube are of course cubes and these truncate first into truncated cubes and finally into cuboctahedra. So we see that the resulting figure has 24 cells of which 16 are tetrahedra and 8 are cuboctahedra. One tetrahedron and two cuboctahedra meet at each edge. The cuboctahedra meet each other at shared square faces and meet the tetrahedra at shared triangular faces. The figure has 32 vertices since the hypercube has 32 edges.

The 16-cell behaves somewhat differently. The 8 vertices become 8 octahedra and the 16 tetrahedral cells also truncate into octahedra. So the resulting figure has 24 octahedral cells and is simply the regular 24-cell. So we have discovered another construction for this curious self-dual figure.

If we now truncate the 24-cell itself, the 24 vertices become 24 cubes whereas the 24 original octahedral cells become 24 cuboctahedra. Thus the figure has 48 cells and 96 vertices. One cube and two cuboctahedra meet at each edge, the cuboctahedra meeting at triangular faces.

So we have discovered two new figures whose cells are mixtures of cuboctahedra with tetrahedra or cubes respectively. But these do not really class as four-dimensional Archimedean figures since a cuboctahedron is not regular.

In a similar way, truncating the 120 cell (whose cells are dodecahedra) results in a figure comprising 120 icosidodecahedra and 600 tetrahedra so that is not Archimedean either. However, truncating the 600-cell (whose cells are tetrahedra and vertex figure is an icosahedron) results in a figure comprising 120 icosahedra and 600 octahedra with one icosahedron and two octahedra meeting at each edge and two icosahedra and five octahedra meeting at each of the 720 vertices. So here is another Archimedean figure.

The various truncations are summarized in the table above. The situation is quite different to that in three dimensions where the cube and octahedron (which are duals) both truncate into cuboctahedra and the dodecahedron and icosahedron both truncate into icosidodecahedra. In three dimensions the truncation process

Object	Cells	Faces	Edges	Verts	Dihedral
Tetroctahedric	10 = 5 tetra + 5 octa	30	30	10	289° 28'
Octicosahedric	720 = 600 octa + 120 icosa	3600	3600	720	357° 07'
Tetricosahedric	144 = 120 tetra + 24 icosa	480	432	96	349° 15' or 346° 54'

generated 7 (of the 13) Archimedean figures whereas in four dimensions it has only revealed two.

In seeking other Archimedean figures observe that since adjacent cells share a face, then all faces must be the same. But the cells must not all be the same otherwise it would be regular. This means that the faces can only be triangles and the cells can only be combinations of tetrahedra, octahedra, and icosahedra. Also there must be at least three cells at each edge just as in three dimensions all solid figures must have at least three faces at each vertex. Another requirement is that the sum of the dihedral angles must not exceed 360° just as the sum of the angles of the faces at a vertex in a three-dimensional figure must not exceed 360°.

We have discovered two, that of 5 tetrahedra and 5 octahedra and that of 120 icosahedra and 600 octahedra. Thorold Gosset (1869–1962), an English lawyer, was the first to enumerate these Archimedean semi-regular figures and as well as these two (which he called the Tetroctahedric and the Octicosahedric), he discovered the only other which he called the Tetricosahedric.

This remarkable figure comprises 120 tetrahedra and 24 icosahedra. It has 96 vertices and these are the points obtained by dividing the 96 edges of the 24-cell in the golden ratio. Three icosahedra and five tetrahedra meet at each vertex. It has 432 edges but it is not that regular since some edges are surrounded by one tetrahedron and two icosahedra and others by three tetrahedra and one icosahedron. Moreover, it is a snub figure and occurs in enantiomorphic forms. Coxeter refers to it as s{3, 4, 3} because of its derivation from the 24-cell.

So in contrast to the 13 Archimedean figures in three dimensions, there are only three such figures in four dimensions; moreover, none is as regular as the cuboctahedron since in every case the faces are not all equivalent.

Note that the honeycomb of octahedra and tetrahedra in three dimensions was classified by Gosset as a four-dimensional figure.

Prisms and hyperprisms

A REGULAR PRISM in three dimensions consists of two regular polygons joined by squares as illustrated in Lecture 2. We can describe the creation of a prism as being done by taking a regular polygon and connecting it point by point to an equal figure in a parallel plane where the distance between the planes is equal to the side of the polygons. Note that a square prism is simply a cube.

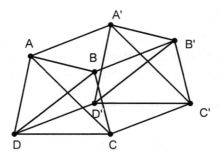

A tetrahedral hyperprism.

We can similarly define a hyperprism as being obtained by taking a polyhedron and joining it to a similar polyhedron in a parallel space. If the polyhedron is a cube then the hyperprism is simply the hypercube. We will now consider some other simple hyperprisms.

If the polyhedron is a tetrahedron, then it is easy to see that the resulting tetrahedral hyperprism shown above must have 8 vertices (4 from each tetrahedron) and 16 edges (6 from each tetrahedron and 4 joining the two tetrahedra). It has 14 faces (4 triangular faces from each tetrahedron plus 6 square faces obtained by joining the 6 pairs of corresponding edges) and 6 cells (the 2 original tetrahedra plus 4 triangular prisms obtained by joining the 4 pairs of corresponding faces of the tetrahedra).

Such a hyperprism is sort of regular in that each vertex is the same – one tetrahedron and three triangular prisms meet at each vertex. On the other hand the edges are not all the same since some such as AB belong to 2 triangular faces (ABC and ABD) and 1 square face ($ABB'A'$) whereas others such as AA' belong to 3 square faces ($AA'B'B$, $AA'C'C$, and $AA'D'D$) and so the tetrahedral hyperprism is not that regular.

We can create a hyperprism starting from a triangular prism itself as shown below. This triangular prismatic hyperprism clearly has 12 vertices (6 from each prism) and 24 edges (9 from each prism and 6 joining the two prisms). It has 19 faces (3 squares and 2 triangles from each prism plus 9 new squares from the 9

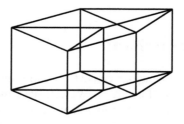

A triangular prismatic hyperprism.

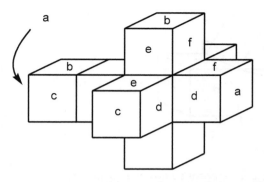 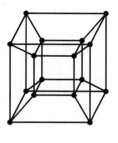

A hypercube as two rings of four cubes.

pairs of edges giving 4 triangles and 15 squares in total) and 7 cells (the 2 original prisms plus 2 new prisms from joining the 2 triangular faces of the original prisms and 3 cubes from joining their square faces giving 4 triangular prisms and 3 cubes in total).

The triangular prismatic hyperprism is again sort of regular in that each vertex is the same – two cubes and two triangular prisms meet at each vertex.

It is always worth checking that the Euler formula $C-F+E-V$ is correctly zero or in other words that $C+E$ and $F+V$ are the same. For the tetrahedral hyperprism they are both 22 and for the triangular prismatic hyperprism they are both 31. Note that the corresponding figure for the hypercube is 40 so in some sense they are simpler structures.

Another approach to understanding these structures is to consider the net for the hypercube which we met in Lecture 3. It is shown above on the left. This net comprises two rings of four cubes wrapped around each other. One ring is the central row of cubes which we know in four dimensions is folded so that the far left face marked "a" is joined to the far right face also marked "a". And the other ring of four cubes consists of the cubes around the central row which we know also meet in four dimensions since for example the two faces marked "e" coincide.

We can easily trace these two rings of cubes in the representation shown above on the right. One ring is composed of the small central cube, the distorted cube above it, the large cube and then the bottom distorted cube. The other ring is composed of the four distorted cubes around the middle. We can also trace two rings another way by taking the small central cube, the distorted cube to the right, the large cube and then the distorted cube to the left as one ring and then the other ring is made of the remaining distorted cubes.

We can construct a four-dimensional figure out of two rings of prisms in many ways. The general case is where there are p q-gonal prisms and q p-gonal prisms such as five hexagonal prisms and six pentagonal prisms. Note that the two types of prisms always meet each other in square faces (remember that all prisms have square faces).

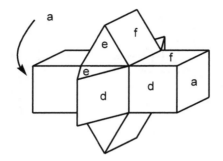

A triangular prismatic hyperprism as two rings of prisms.

For example, suppose we have a row of three cubes and then around the central cube place four triangular prisms as shown above. Again we can identify the pairs of faces and we have a ring of three cubes and a ring of four triangular prisms wrapped around each other. Indeed, this is the triangular prismatic hyperprism once more. The cubes meet the triangular prisms and each other in square faces whereas the triangular prisms meet each other in triangular faces.

A particularly nice example is that of two rings of three triangular prisms. In this case the figure comprises 6 cells all of which are triangular prisms and so it has a high degree of regularity. It has 9 vertices, 6 triangular faces and 9 square faces and 18 edges. All the vertices are the same and 4 cells meet at each vertex. Moreover, all the edges are the same and belong to 2 square faces and 1 triangular face. We can call it a triangular pseudohyperprism – it's not properly a hyperprism since a hyperprism is obtained by joining two identical polyhedra by parallel lines. We find that $C+E = F+V = 24$.

Another fairly simple example of a true hyperprism is the square pyramidal hyperprism obtained by joining two square pyramids. It has 10 vertices (5 from each pyramid), 21 edges (8 from each pyramid and 5 joining the two pyramids), 18 faces (4 triangles and 1 square from each pyramid plus 8 square faces from joining the 8 pairs of corresponding edges giving 8 triangles and 10 squares in total) and finally 7 cells (the 2 original pyramids plus 1 cube obtained by joining the square bases of the pyramids and 4 triangular prisms obtained by joining the four triangular faces of the pyramids). We check that $C+E = F+V = 28$. This figure is not at all regular since not even the vertices are the same (that's largely because the vertices of a square pyramid are not all the same in the first place). Moreover, it has three different kinds of cells.

Although a square pyramid is not regular, two pyramids form an octahedron which is regular, so we should consider the octahedral hyperprism. It has 12 vertices (2 from each octahedron), 30 edges (12 from each octahedron plus 6 joining the two octahedra), 28 faces (8 triangles from each octahedron plus 12 squares from joining the 12 pairs of corresponding edges) and finally 10 cells (the 2 original octahedra plus 8 triangular prisms obtained by joining the 8 faces of the octahedra). Note that $C+E = F+V = 40$, the same as the hypercube.

C More on Four

Object	Cells, C	Faces, F	E	V	$C+E$
4-simplex	5 = 5 tetrahedra	10 = 10t	10	5	15
Squ pyrd hypyrd	6 = 2 pyr, 4 tetra	13 = 12t+1s	13	6	19
Tri prism hypyrd	6 = 1 triprsm, 3 pyr, 2 tetra	14 = 11t+3s	15	7	21
Tetra hyprism	6 = 2 tetra, 4 triprisms	14 = 8t+6s	16	8	22
Tri pseudohyprism	6 = 6 triprisms	15 = 6t+9s	18	9	24
Cubic hypyrd	7 = 1 cube, 6 pyr	18 = 12t+6s	20	9	27
Squ pyrd hyprism	7 = 2 pyr, 1 cube, 4 triprsm	18 = 8t+10s	21	10	28
Tri prsm hyprism	7 = 4 triprisms, 3 cubes	19 = 4t+15s	24	12	31
Hypercube	8 = 8 cubes	24 = 24s	32	16	40
Octa hyprism	10 = 2 octa, 8 triprisms	28 = 16t+12s	30	12	40
Tetroctahedric	10 = 5 tetra, 5 octa	30 = 30t	30	10	40
16-cell	16 = 16 tetrahedra	32 = 32t	24	8	40

We conclude with a few words about hyperpyramids. A pyramid is obtained by joining a point to each vertex of a plane figure (a tetrahedron is simply a triangular pyramid). A hyperpyramid is similarly obtained from a polyhedron by joining each vertex to a point in the fourth direction. If the polyhedron is a tetrahedron then we get the 4-simplex (or 5-cell) and so the 4-simplex is simply a tetrahedral hyperpyramid.

If we start with a cube we get a cubic hyperpyramid. This has 9 vertices (8+1), 20 edges (12+8), 18 faces (6 squares plus 12 triangles) and 7 cells (1 cube plus 6 square pyramids). So $C+E = F+V = 27$.

If we start with a triangular prism we get a rather odd figure. It has 7 vertices (6+1), 15 edges (9+6), 14 faces (3 squares plus 2+9 triangles) and 6 cells (1 triangular prism, 3 square pyramids and 2 tetrahedra (triangular pyramids)). So $C+E = F+V = 21$.

And if we start with a square pyramid we get a square pyramidal hyperpyramid. This has 6 vertices (5+1), 13 edges (8+5), 13 faces (1 square plus 4+8 triangles) and 6 cells (2 square pyramids and 4 tetrahedra). So $C+E = F+V = 19$. Note that the two square pyramids are similarly arranged. We can think of this figure as being obtained from a base square in two dimensions and then we add two further points one in each of two other dimensions and join these points to the square and to each other.

All the figures we have discussed in this section plus the tetroctahedric are summarized in the table above in ascending order of the values of $C+E$. Note that all the faces are either triangles or squares.

Apart from the truly regular figures, two are of especial note for their regularity. One is the tetroctahedric which is notable in that all its faces are triangles. The other is what we have called the triangular pseudohyperprism all of whose cells are the same even though the cells themselves being triangular prisms are not regular.

D Schlegel Images

IN LECTURE 5 we introduced the idea of a Schlegel diagram as a convenient way of representing the topology of a three-dimensional object such as an octahedron as a two-dimensional image.

In this appendix we look at this process for three-dimensional objects in a little more detail and then consider the corresponding process for four-dimensional objects.

Schlegel diagrams

A SCHLEGEL DIAGRAM of a three-dimensional object is obtained by projecting the object onto a plane. If the viewpoint is sufficiently close to one face then the projection of that face surrounds the projection of the other faces. This means that in topological terms we can think of the space outside the image as representing the near face.

Take the cube for example (a bit of a special case). If we view a cube of edge $2x$ from a point $2x$ from the centre on an axis through the centre of a face then we see the second image below. If we move further away to say $5x$ then the opposite face becomes proportionately much larger. If we move to infinity then we get an orthogonal projection and the near and far faces coincide so all we see is a square. However, we might consider the view from the circumsphere as canonical and in the case of the cube this is at distance $\sqrt{3}x$ ($= 1.732x$) from the centre. This is the first image below and is the one used in Lecture 5.

Note that in each case the projections are onto the plane through the centre of the cube so as we move away the near face gets smaller but the far face gets a bit larger.

The images of the various squares are of three different kinds. First there is the image of the opposite face at level 1, then the four side faces at level 2 and finally the near face at level 3. We can designate the pattern as (1, 4, 1).

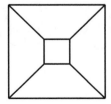 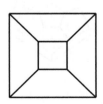

Projections of a cube from √3x, 2x, 5x, and infinity.

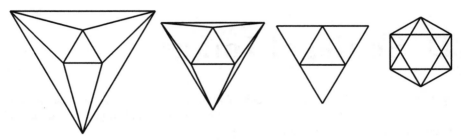

Projections of an octahedron from √2x, 2x, √6x, and infinity.

We can do the same sort of thing with the octahedron. One key difference is that the opposite faces of an octahedron are oriented differently and so do not coincide when viewed from infinity. Another big difference is that if we view an octahedron from far away then we can see some of the side faces whereas in the case of the cube that is not possible if we look at it face on. So if we get too far away the projection of the nearest face no longer surrounds the projection of all the other faces. If the edge of the octahedron is $2x$ then the critical point is when we view it at a distance of $\sqrt{6}x$ (= 2.449x) from the centre. Three of the triangles in the projected image then degenerate.

The diagrams above show the view from the circumsphere (1.414x), from $2x$, from the critical point (2.449x), and from infinity.

The projection of the octahedron has triangles at four different levels and the overall pattern is (1, 3, 3, 1).

The dodecahedron portrays similar behaviour to the octahedron. Again the opposite faces are oriented differently and again if viewed from afar then some of the side faces are visible. Suppose that the edge of the dodecahedron is $2x/\tau$ where τ is the golden number. Then the circumradius is $\sqrt{3}x$ (= 1.732x) and the critical point where the projected image of the first set of faces degenerates is at $\sqrt{(4\tau+3)}x$ (= 3.078x) from the centre.

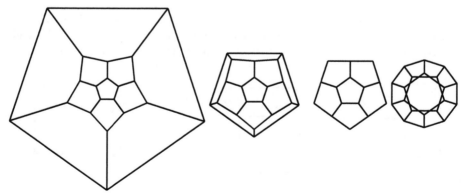

Projections of a dodecahedron from √3x, 2.5x, √(4τ+3)x, and infinity.

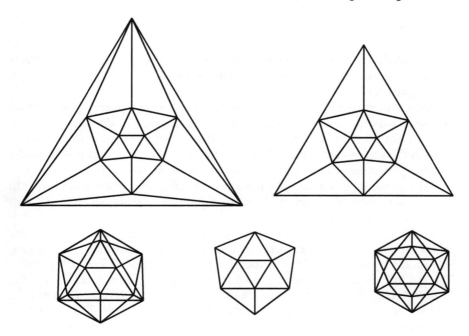

Projections of an icosahedron from $\sqrt{(3-\tau)}x$, $\sqrt{(3/5)}\tau x$, $2x$, $\sqrt{3}\tau x$, and infinity.

The diagrams opposite show the view from the circumsphere ($1.732x$), from $2.5x$, from the critical point ($3.078x$), and from infinity.

The projection of the dodecahedron has pentagons at four different levels and the overall pattern is (1, 5, 5, 1).

The tetrahedron is very boring. If you look at it face on then no matter how far away or near the projection is always just a triangle subdivided into three equal triangles. The pattern is simply (3, 1).

To complete the five Platonic solids we need to consider the icosahedron which behaves somewhat differently. From afar, two distinct sets of side faces are visible and the projections of these sets degenerate at different points. So there are two critical points.

If the edge is $2x/\tau$ (we use the same scale as for the dodecahedron since the figures are dual), then the circumradius is $\sqrt{(3-\tau)}x = 1.175x$. The first critical point is at $\sqrt{(3/5)}\tau x = 1.253x$ and the second critical point is at $\sqrt{3}\tau x = 2.802x$.

The diagrams above show the view from the circumsphere ($1.175x$), from the first critical point ($1.253x$), from $2x$, from the second critical point ($2.802x$), and from infinity. The icosahedron has triangles at six different levels and the overall pattern is (1, 3, 6, 6, 3, 1).

Before embarking on projections of four-dimensional figures, it is worth considering whether the projection of the three-dimensional figures would help a Flatlander to gain a better impression of them. Well, maybe. It's hard to know.

The hypercube

THE OBVIOUS ANALOGY of the Schlegel diagrams for three-dimensional objects is to consider the projection of a four-dimensional object into (or onto) a three-dimensional space.

Again there are a number of possibilities. We can project from various distances and from various angles. The natural analogy to the projections we have just seen is to project from a point on the circumhypersphere where an axis through the centre passes through the centre of a cell. We might then consider how the projection changes as the viewpoint moves along the axis to infinity.

The projected images are of course three-dimensional figures and in order to gain a good comprehension of them the figures are rotated slightly and also presented as stereo images.

We begin with the hypercube. If the edge of the hypercube is $2x$ then the radius of the circumhypersphere is also $2x$. That is obtained by applying Pythagoras in four dimensions so the square of the radius is $x^2+x^2+x^2+x^2 = 4x^2$. The result is the familiar figure shown below. The inner cube is at level 1, the distorted cubes around it are level 2 and the outer cube is level 3. So using the notation of the previous section, the image has pattern (1, 6, 1). If we move the point of projection to infinity then the inner and outer cubes coincide and the image is uninteresting just as the projection of a cube onto a plane from infinity is uninteresting.

We can of course project from points in different directions. Obvious choices are through the centre of a face, through the centre of an edge and through a vertex. And we can consider projections from the circumhypersphere or from infinity or somewhere in between.

Suppose for example that we project through the centre of an edge from the circumhypersphere. This gives the top image opposite which is somewhat confusing. There are eight cubes here but only three somewhat distorted ones are clearly visible.

Things become a bit clearer if the viewpoint is moved to infinity as shown next. However, there is still some confusion because two edges (and hence two

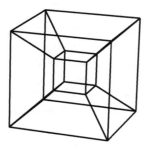 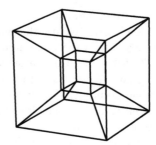

Projection of a hypercube cell first from the circumhypersphere.

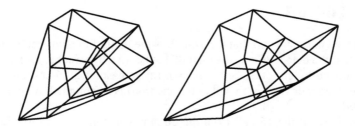

Projection of a hypercube edge first from the circumhypersphere.

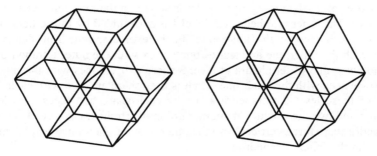

Projection of a hypercube edge first from infinity.

pairs of points) have become exactly superimposed and appear as a single edge in the middle of the diagram so that it looks overall like a hexagonal prism.

If we move the direction of projection very slightly then the two edges become distinct and we get the view shown below. It is this view that was used in Lecture 3 when we were discussing the appearance of a hypercube crossing our space in various orientations. It should now be clear that there are three cubes around both of the now separated central edges. The other two cubes are extremely flattened in this view and comprise the two hexagonal faces of the now broken prism.

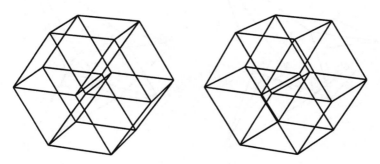

Projection of a hypercube nearly edge first from infinity.

The 16-cell

IT WILL BE REMEMBERED that the 16-cell is the dual of the hypercube in the same way that the octahedron is the dual of a cube. In a hypercube, each cell (a cube) is surrounded by six other cells (cubes). In the case of a 16-cell, the individual cells are tetrahedra and each cell (a tetrahedron) is surrounded by four other cells (tetrahedra).

The view of the 16-cell from the circumhypersphere in a direction through the centre of a cell is shown below. The overall figure is a tetrahedron in blue (the cell nearest the view point). In the centre is a smaller tetrahedron in red (this is the opposite cell to the view point). A dot is shown in their common centre to aid perspective. The inner tetrahedron at level 1 is surrounded by four others at level 2, one on each face of the inner one. Each level 2 tetrahedron shares three vertices with the inner one and shares one vertex with the outer tetrahedron.

Another group of tetrahedra at level 3 share two vertices with the inner one and two vertices with the outer one – that is they share an edge with each. There are six of these level 3 tetrahedra and they each share two faces with level 2 tetrahedra and two faces with yet another group of tetrahedra at level 4. The level 4 tetrahedra share three vertices with the outermost tetrahedron at level 5 and one vertex with the inner tetrahedron.

So there are five levels for the projection of the 16-cell and the symbol for the pattern is (1, 4, 6, 4, 1). These total 16 as expected.

It should be noted that the tetrahedra at level 4 are in fact degenerate in this view and have fallen flat. So (rather surprisingly) being on the circumhypersphere is not close enough in four dimensions to ensure that all cells are projected properly inside the image of the outer cell. Using the terminology for discussing the projections of three-dimensional objects we see that the point on the circumhypersphere is in fact a critical point and we should really project from a little closer.

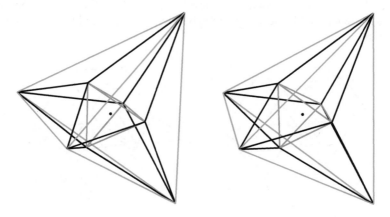

Projection of a 16-cell, cell first, from the circumhypersphere.

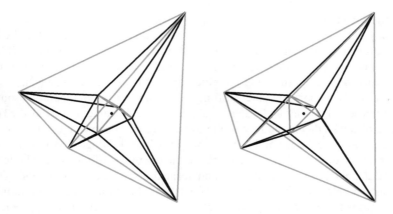

Projection of a 16-cell, cell first, from 0.75x.

If the edge of the 16-cell is $\sqrt{2}x$ then the circumradius is x and the centre of a cell is at $0.5x$ – this is the radius of the inscribed hypersphere. So we consider moving the viewpoint somewhere between x and $0.5x$. In the following diagrams the projection is onto the space of the near cell itself. This means that the image of the near cell remains constant.

The view above is from $0.75x$. The level 4 tetrahedra are now no longer degenerate. This image is quite attractive and physical models of it are instructive. Such models were commercially available in Germany as educational aids/toys in the period from about 1890 to 1910.

If we now move the projection point away towards infinity then the inner tetrahedron grows. Its vertices reach the faces of the outer tetrahedron when the viewpoint is at x on the circumhypersphere as seen earlier.

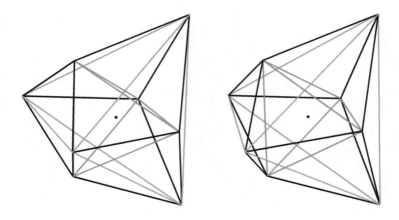

Projection of a 16-cell, cell first, from 2x.

As the viewpoint moves further away, the vertices of the inner tetrahedron project beyond the faces of the outer tetrahedron and so the level 4 tetrahedra become reversed. The image when the viewpoint is at 2*x* was shown above. It has a slight appearance of being a distorted cube. And note how the red edges of the inner tetrahedron are approaching the blue edges of the outer tetrahedron.

When the viewpoint reaches infinity, the edges of the inner and outer tetrahedra meet. The inner and outer tetrahedra in fact become related as the two tetrahedra of the stella octangula and the overall figure becomes the surrounding cube. Note that the cube is composed of the black lines joining the original inner and outer tetrahedra and those tetrahedra are the diagonals of the cube. The level 3 tetrahedra then become degenerate and form the six faces of the cube. This view is shown below and was the view of the 16-cell used in Lecture 3 although the latter was rotated differently. This figure seems a little difficult to appreciate. We are so used to seeing cubes that the tetrahedra are obscured. Moreover, the cube is presented at a slightly unfamiliar angle.

Does this sequence of views of the 16-cell help us humble three-dimensional beings to appreciate its construction? The author thinks so although the reader might disagree.

Another way to see what is going on is to consider the image starting from the inner tetrahedron as successive levels are added. In the stereo images opposite, of the view from 0.75*x*, the hidden lines are removed and concave edges are shown in thinner lines.

First we start with the level 1 inner tetrahedron. Then we add the four level 2 tetrahedra which hide the level 1 tetrahedron completely.

We next add the six level 3 tetrahedra which in turn hide the level 2 tetrahedra. And finally, we add the four level 4 tetrahedra which hide the level 3 tetrahedra and this results in the overall shape of the outer level 5 tetrahedron.

So the overall structure is a bit like a Russian doll with each layer of tetrahedra completely covering the others in turn.

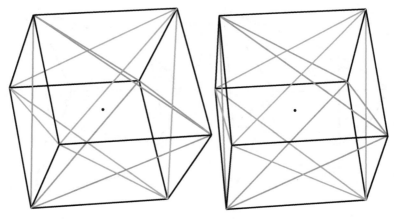

Projection of a 16-cell, cell first, from infinity.

D Schlegel Images

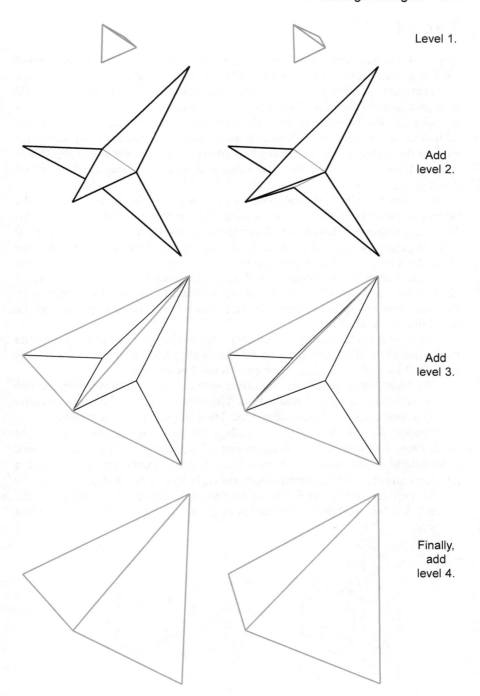

Level 1.

Add level 2.

Add level 3.

Finally, add level 4.

Building up the image layer by layer.

The 24-cell

WITH a certain amount of trepidation we can consider similar projections of the other regular four-dimensional figures such as the 24-cell.

There are a number of ways of constructing a 24-cell. One is to add hyperpyramids to the cells of a hypercube (this was described in Lecture 3 when considering honeycombs in four dimensions). Another way is to join the midpoints of the edges of a 16-cell as mentioned in Appendix C. By duality, if we join the midpoints of the faces of a hypercube we also get a 24-cell. This is because the hypercube and 16-cell are dual and edges and faces are dual in four dimensions.

We start by joining the midpoints of the edges of the 16-cell and use the second of the views in the previous section with the viewpoint obtained at $0.75x$. The result is shown below. It is a bit confusing at first sight. Remember that a 24-cell comprises 24 octahedra. There is an inner octahedron in red and an outer octahedron in green. These correspond to truncating the inner tetrahedron and outer tetrahedron of the parent 16-cell. Lying between these two octahedra is a black cuboctahedron. The cuboctahedron is joined to the inner octahedron by pale blue lines and to the outer one by pale brown lines. We now need to find the remaining 22 octahedra.

There is a level 2 octahedron on each face of the inner red one. Its opposite face is one of the triangular faces of the black cuboctahedron and these two faces are joined by blue edges. There are eight level 2 octahedra.

The level 3 octahedra share one vertex with the inner red octahedron and one vertex with the outer green octahedron. The square faces of the cuboctahedron are equatorial squares of these octahedra. There are six level 3 octahedra.

Finally, there is a level 4 octahedron on each face of the outer green octahedron. Its opposite face is again one of the triangular faces of the black cuboctahedron. This means that each level 4 octahedron shares a face with a level 2 octahedron. And naturally there are eight level 4 octahedra.

So there are five levels for the projection of the 24-cell (the same as for the 16-cell) and the symbol for the pattern is (1, 8, 6, 8, 1) and these total 24 as required.

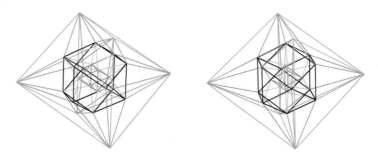

Projection of a 24-cell, cell first, from 0.75x.

D Schlegel Images

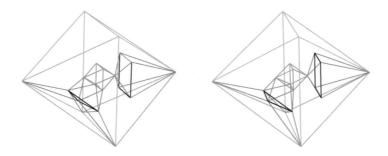

The projection showing just one cell from each level.

The image above shows just one octahedron from each level and might help. On the right is a level 3 octahedron joining the vertices of the inner and outer octahedra. At the bottom left a level 2 and a level 4 octahedron are shown – they share a common face and a face of the inner and outer octahedra.

Further insight is gained by considering the images as the various levels are added. In the images below the hidden lines are removed and concave edges are thinner.

We start with the level 1 inner octahedron. Adding the eight level 2 octahedra produces the cuboctahedron with deep dimples on its square faces reaching down to the vertices of the hidden inner octahedron. Adding the six level 3 octahedra fills these dimples and produces a figure which reaches out to the final vertices of the full image. Note that the triangular faces of the cuboctahedron are still visible – they are of course actually faces of the level 2

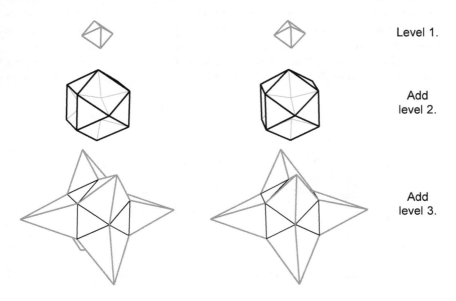

Level 1.

Add level 2.

Add level 3.

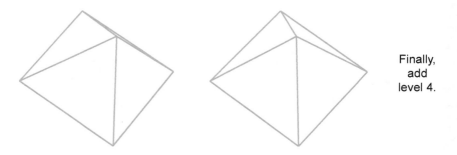

Finally, add level 4.

octahedra. And finally, adding the very flattened level 4 octahedra completes the outer level 5 octahedron.

Note that the cuboctahedron is a cross-section of the 24-cell. There are in fact 12 such cuboctahedra, one for each pair of opposite cells. Another is shown below in thicker lines. It is so distorted that it has collapsed into a single plane and so appears as a Schlegel diagram of a cuboctahedron.

The 24 vertices of a 24-cell define three associated hypercubes and three associated 16-cells.

The hypercubes are constructed from the equatorial squares of the octahedra. Consider for example the level 3 octahedron on the right in the earlier diagram and its equatorial square that is a face of the cuboctahedron. This square shares edges with equatorial squares of all the adjacent octahedra. If we trace these adjacent squares throughout the 24-cell we find that we have discovered a hypercube lurking inside the 24-cell. The hypercube has 24 faces, one for each octahedron of the 24-cell.

However, we can do this process by starting with any of the equatorial squares of the first octahedron. And since an octahedron has three equatorial squares it follows that there are three related hypercubes inscribed in a 24-cell. The three are shown above opposite using the same colours as the 24-cell.

It will be recalled that compounds can be formed in three dimensions such as the compound of five cubes. If we put the three hypercubes associated with a

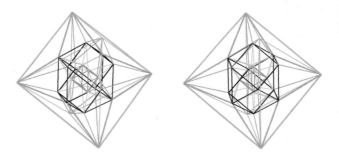

Another cuboctahedron appearing as a Schlegel diagram.

D Schlegel Images   283

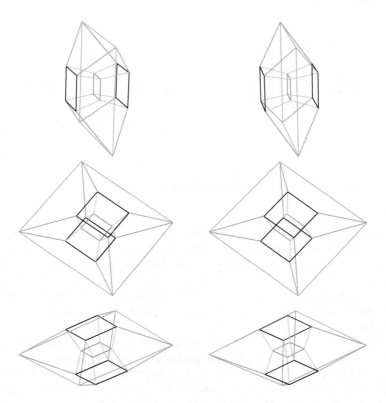

Three hypercubes related to a 24-cell.

24-cell together then we form a compound of three hypercubes. This is shown below where the hypercubes are now red, blue and green.

The compound looks remarkably like the original 24-cell with the colours of the edges changed. But this is not really the case. The faces of the 24-cell are all

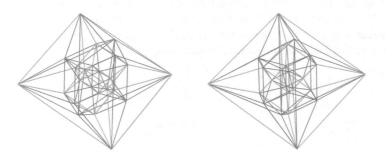

The compound of three hypercubes.

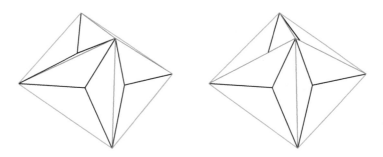

The compound of three hypercubes with hidden lines removed.

triangles whereas those of the hypercubes are of course squares. Thus although the edges are all the same the faces are not and so the two structures are quite different. This difference is clarified by removing the hidden lines. In the case of the 24-cell all that can be seen is the outer level 5 octahedron as shown earlier. However, in the case of the compound of hypercubes we get the dimpled figure shown above where the triangular faces are parts of the square faces of the hypercubes.

Each hypercube shares 16 vertices with the 24-cell. The remaining 8 vertices define a related 16-cell. This is to be expected because the hypercube and 16-cell are dual and the 24-cell is self-dual – as a consequence any relationship between the 24-cell and a hypercube must be mirrored by a similar relationship with a 16-cell. The 24 edges of the 16-cell are those diagonals of the octahedra which join the pairs of points not on the equatorial squares forming part of the corresponding hypercube. In other words, starting with a diagonal of an octahedron we build up the 16-cell using adjacent diagonals just as the hypercube was built up using adjacent equatorial squares. And since an octahedron has three diagonals it follows that there are three associated 16-cells just as there are three hypercubes.

The 16-cell corresponding to the first of the hypercubes is shown below. Note that the edges of the 16-cell are not edges of the 24-cell. This corresponds to the fact the the faces of the hypercube are not faces of the 24-cell either. Remember that in four dimensions the dual of a line is a plane and vice versa whereas in three dimensions a line is self-dual.

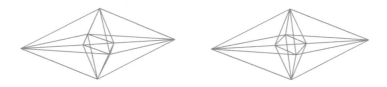

A 16-cell from 8 points of a 24-cell.

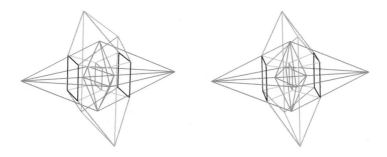

A compound of a hypercube and corresponding 16-cell.

We can merge the 16-cell and corresponding hypercube to produce a compound figure using all 24 vertices of the original 24-cell. And naturally there is also a compound of three 16-cells.

In three dimensions we met such compounds as the stella octangula (two tetrahedra in a cube), the mixed compound of cube + octahedron and the compound of five cubes. We have just seen that in four dimensions there are compounds of three 16-cells, three hypercubes, and a mixed compound of hypercube + 16-cell. There are others as well but enough is enough.

The 120-cell, 600-cell and tetroctahedric

SIMILAR projections and images can be created for the 120-cell (whose cells are dodecahedra) and the dual 600-cell (whose cells are tetrahedra). Needless to say they are not easy to depict in the same manner because there are just too many edges (1200 and 720 respectively). However, it is worth stating the general arrangement.

The projection of the 120-cell has a level 1 inner dodecahedron at its centre. This is surrounded by 12 level 2 dodecahedra one on each face of the inner dodecahedron. The result has 20 dimples and the level 3 dodecahedra lie in these dimples. And so on. The final pattern is (1, 12, 20, 12, 30, 12, 20, 12, 1) which adds to 120. So the dodecahedra lie in nine different levels.

The 600-cell is even more hazardous. It starts with a level 1 inner tetrahedron at its centre and this is surrounded by 4 level 2 tetrahedra. The level 2 tetrahedra now have 12 exposed faces (3 each) and on these are placed 12 level 3 tetrahedra. And so on. Altogether the pattern has an amazing 31 levels: (1, 4, 12, 24, 12, 4, 24, 24, 32, 24, 12, 24, 28, 24, 24, 54, 24, 24, 28, 24, 12, 24, 32, 24, 24, 4, 12, 24, 12, 4, 1).

Although we cannot depict these elaborate objects we will finish by showing the tetroctahedric introduced in Appendix C. This is the simplest of the four-

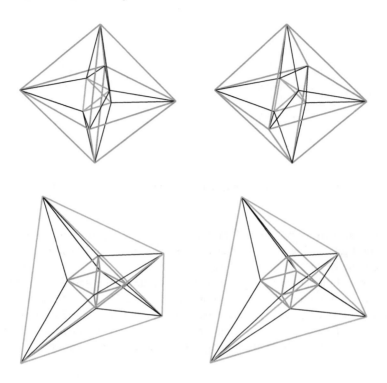

Two images of the tetroctahedric.

dimensional Archimedean figures and has 10 cells, 5 tetrahedra and 5 octahedra. It is thus possible to produce two different images, one with an octahedron at the centre and the other with a tetrahedron at the centre. Both images are shown above.

They are intriguing images with a mixture of tetrahedra and octahedra very clearly visible. They echo the structure of the 16-cell and the octahedra give a flavour of the 24-cell without being overwhelmingly complex. In the top view above, the inner tetrahedron at the centre at level 1 is surrounded by four octahedra at level 2. Then there are four tetrahedra at level 3 and finally the outer octahedron at level 4. So the pattern is (1, 4, 4, 1).

The reader is invited to contemplate images of some of the other simple figures discussed in Appendix C such as the triangular pseudohyperprism which is composed of six triangular prisms.

E Crystals

A SET OF LECTURES entitled gems of geometry must clearly say something about crystals. We start by considering the problem of packing spheres.

Packing of spheres

THE PROBLEM of how best to pack spheres together to occupy the least space has been studied for a long time. Familiar examples of this occur in the packing of oranges on a greengrocer's display or the stacking of cannonballs in a pile.

Let us first consider the analogous simple problem of packing circles in two dimensions, that is on a plane. We could of course pack them in rows and columns but this is not at all good. Since the area of a circle is πr^2 and the area of a square of side $2r$ is $4r^2$ the fraction of the space occupied by the circles is $\pi r^2 \div 4r^2 = \pi/4$ or about 0.785....

A much better packing is where the centres of the circles are arranged as the vertices of the tiles in the triangular tiling. The area of a triangle of side $2r$ is $\sqrt{3}r^2$ but this corresponds to the space occupied by only half a circle (three sectors each of which is one-sixth of a circle) so the fraction of the space occupied by the circles is $\frac{1}{2}\pi r^2 \div \sqrt{3}r^2 = \pi/2\sqrt{3}$ or about 0.906.... This is in fact the best

Area of square is 2r × 2r = 4r²,

area of 4 sectors of circle is πr^2,

ratio = $\pi/4$ = 0.785....

Packing circles as in a square tiling is very inefficient. Only some 78.5% of the space is utilized.

Area of triangle is r × √3r = √3r²,

area of 3 sectors of circle is $\frac{1}{2}\pi r^2$,

ratio = $\pi/2\sqrt{3}$ = 0.906....

Packing circles as in a triangular tiling is much more efficient. Over 90.6% of the space is utilized.

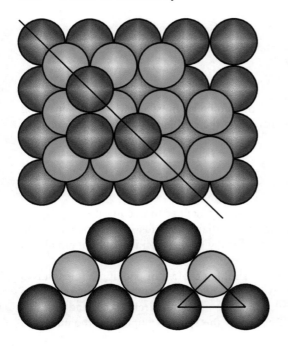

Alternate layers of red and black cannonballs where each layer is packed as a square tiling.

About 74% of the space is utilized.

A vertical and diagonal slice at 45° through the points of contact between the two layers.

The base of the triangle is $2\sqrt{2}r$ and its height is $\sqrt{2}r$.

packing possible. We recall from Lecture 2 that trees in an orchard are grown arranged as in the triangular tiling because this gives maximum land utilization.

The analogous problem in three dimensions is how to pack spheres so that the least space is occupied.

Cannonballs are usually piled with the first layer of balls arranged in the square pattern. Balls for the second layer are then placed above the gaps so that each ball in the second layer touches four balls in the layer below.

In the diagram above the first layer of balls is shown in black and the second layer is shown in red. (We are looking down of course.) Each red ball touches four black balls in the layer below and four other red balls in its own layer. We can then add a third layer of black balls in the gaps in the second layer so that the balls in the third layer are immediately above those in the first layer and so on. It is clear that each ball then touches twelve other balls.

In order to compute the packing consider a section downwards and at 45° so that it goes through the points where the black and red balls touch. In this section the distance between the centres of the black balls on the bottom layer (we assume the balls have radius r) is $2\sqrt{2}r$. By considering the triangle of centres between two black balls and the red ball above we have a triangle whose base is $2\sqrt{2}r$ and whose sloping sides are $2r$. The height of the triangle is therefore $\sqrt{2}r$. So we have shown that the layer of red balls is $\sqrt{2}r$ above the layer of black balls whereas if the red balls had been directly over the black balls the red layer would have been $2r$ above the black layer – an improvement of a factor of $\sqrt{2}$. Incidentally this slice reveals the square arrangement once more.

E Crystals

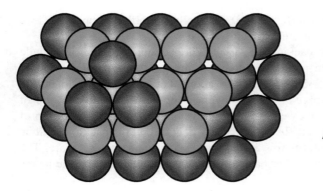

Alternate layers of red and black cannonballs where each layer is packed as a triangular tiling.

Again 74% of the space is utilized.

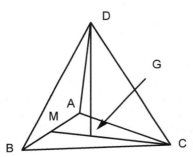

Using Pythagoras on AMC we deduce that MC is √3r.

Hence GC is 2√3r/3.

Then in DGC we apply Pythagoras again and obtain DG = √(8/3)r.

Now since the volume of a sphere is $\frac{4}{3}\pi r^3$ and the volume of a cube of side $2r$ is $8r^3$, the density of packing for a simple cubic arrangement would be $\frac{4}{3}\pi r^3 \div 8r^3 = \pi/6$ or about 0.523.... It follows that the density for the arrangement we have been discussing is $\sqrt{2}\pi/6$ or about 0.740....

Another way to stack things is to start with the triangular pattern for the first layer and then to place the second layer such that each ball is above a gap in the first layer. This time however each ball in the second layer will only touch three balls in the layer below. Moreover, the balls in the second layer are directly above only half of the gaps in the first layer. If a third layer is now added then we have a choice – either the balls in that layer are above the spare gaps in the first layer or they can be above the first layer. It clearly doesn't make any difference to the packing density. We choose to place the third layer over the spare gaps in the first layer. The fourth layer can then be directly above the gaps in the second layer and so on. Observe that although each red ball only touches three balls in the layer above and in the layer below, it now touches six other red balls in its own layer. So altogether each ball still touches twelve other balls.

In order to compute the packing we note that the centres of a red ball and the three black balls in the layer below are arranged as the vertices of a tetrahedron $ABCD$. If an equilateral triangle ABC has side $2r$ then its height MC is $\sqrt{3}r$. Since the centre of a triangle G is one-third of the way up the median MC this means that the distance between a vertex C at the foot of the tetrahedron and the point

G of the base under the top vertex is $2\sqrt{3}r/3$. Finally, from the triangle DGC we deduce that the height DG of the tetrahedron is $\sqrt{(8/3)}r$.

Now the density of the packing for a triangular packing of spheres in a single layer can be calculated in much the same way as for the triangular tiling of circles. We get $^2/_3\pi r^3 \div 2\sqrt{3}r^3 = \pi/3\sqrt{3}$ or about 0.604.... This would be the packing density if the triangular packed layers were placed directly on top of each other. But we know that we have an improvement over that by the ratio of the height of the tetrahedron to $2r$. So we finally deduce that the packing must be $\pi/3\sqrt{3} \times 2/\sqrt{(8/3)}$. This reduces to $\sqrt{2}\pi/6$. Gosh, that's the same as before. The saving by making the layers better packed is exactly compensated by the layers being further apart.

The reason why the two packings give the same result is that *they are the same packing* but looked at from a different angle. In order to justify this perhaps surprising statement we will show that both configurations have planes revealing the square and triangular tilings.

If we make a square pyramid on a base using the square packing as below then the four sloping sides reveal the triangular tiling where each sphere is surrounded by six others. Moreover, we have already seen that a vertical slice through a diagonal reveals the square tiling again. Since there are two diagonals this means that three directions of planes in total show the square tiling and four planes show the triangular tiling – seven planes in total.

If we make a hexagonal pyramid as opposite in which the horizontal layers have the triangular packing then three of the six sloping faces reveal the square packing and the other three reveal the triangular packing. Again we have seven planes – four show the triangular packing and three show the square packing.

So it is clear that the packings are the same. Moreover, we recall from Lecture 2 that a cuboctahedron has 12 vertices and 14 faces arranged as 7 parallel pairs. It is now easy to see that the 12 spheres touching a sphere are arranged as

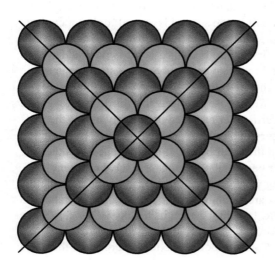

The four sloping sides show the triangular packing.

The horizontal planes and two vertical diagonal planes show the square packing.

3 square + 4 triangular.

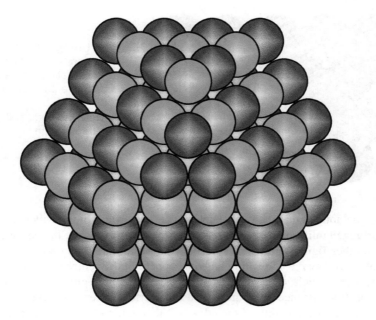

The north, south-east and south-west sloping sides plus the horizontal planes show the triangular packing. The north-west, north-east and south sloping sides show the square packing. 3 square + 4 triangular.

the vertices of a cuboctahedron. The triangular faces of the cuboctahedron correspond to the four planes with the triangular packing and the square faces of the cuboctahedron correspond to the three planes with the square packing.

Another way of looking at this is to recall that the dual of the cuboctahedron is the rhombic dodecahedron which has 14 vertices and 12 faces. We can place a sphere inside a rhombic dodecahedron so that the 12 faces are all tangent planes to the sphere. The sphere touches each face at its centre. Now a honeycomb can be formed from rhombic dodecahedra and if this is done then the spheres inside will all be touching each other and form the arrangement we have been discussing. So it all fits together.

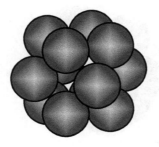

The 12 spheres surrounding one sphere are arranged as the vertices of a cuboctahedron.

292   Gems of Geometry

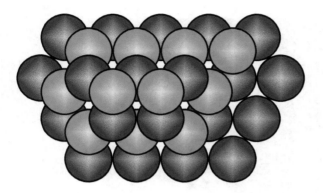

Hexagonal close packing in which alternate layers are above each other.

Note the vertical gaps through the layers.

It will be noticed that the twelve spheres at the vertices of the cuboctahedron are not arranged uniformly. There are big gaps between those at the corners of a square. Another figure with twelve vertices is the icosahedron and if we place spheres at its vertices just touching an equal sphere in the centre then none of the twelve spheres touch each other. This is because the circumradius of an icosahedron (the distance from the centre to a vertex) is less than the length of the edge. In fact if the length of the edge is 2 then the circumradius is $\tau\sqrt{(3-\tau)}$ = 1.902... where τ is the golden number.

Because of this apparent slack it was wondered for many centuries whether a better packing of spheres could be found. Kepler conjectured that the packing we have described was the best but the matter was not settled until 1998 when Tom Hales showed conclusively with a computer assisted proof that this traditional packing could not be bettered.

We conclude by considering the other way of arranging the triangular layers. If we place the third layer directly over the first layer and then the fourth directly over the second and so on, we get the arrangement above. This is called the hexagonal close packing. It has the same packing density of course. However, there are a number of differences. For one thing there are vertical gaps all the way through. And although each sphere touches 12 others they are no longer arranged as the vertices of a cuboctahedron but as the vertices of a related figure which consists of a cuboctahedron with one "hemisphere" twisted in relation to the other.

 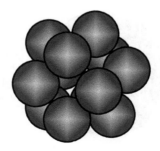

In the hexagonal close packing the 12 spheres surrounding one sphere are arranged as the vertices of a twisted cuboctahedron.

Crystals

THE PACKING OF SPHERES naturally leads us to consider the arrangement of atoms in crystals. This is a complex subject and we cannot cover it thoroughly. However, since these lecture notes are entitled *Gems of Geometry* it would seem unreasonable not at least to consider the arrangement in some compounds such as those constituting gemstones.

The arrangement of crystals is primarily driven by considerations of the minimization of energy; not only does this involve the distance between atoms but complexity is introduced because of valency and bond angles.

One very obvious possibility is where the atoms are placed at the vertices of a honeycomb of cubes. This is not a particularly dense packing. If we have touching spheres of radius r at each point then this means that each cubic cell has side $2r$ and thus volume $8r^3$. Each cell contains one-eighth of each of eight spheres and thus the total volume is $4/3 \pi r^3$ so that the proportion occupied by the spheres is $4/3 \pi r^3 \div 8r^3 = \pi/6$ or about 0.523.... So this simple cubic lattice has a packing of 52% compared with the maximal 74% discussed above.

Another common structure is where the basic cubic cell also has an atom at the centre of the cube. For a cube of side $2r$ the space diagonal is $2\sqrt{3}r$. For equal spheres at the centre and vertices this means that touching spheres have to have radius $1/2\sqrt{3}r$; the volume of each sphere is then $1/2\sqrt{3}\pi r^3$. Note that the sphere at the centre touches the spheres at the vertices but those at the vertices no longer touch each other. Each cell now contains one-eighth of each of eight spheres plus one sphere at the centre – total two spheres and so the total volume is $\sqrt{3}\pi r^3$. It follows that the proportion occupied by the spheres is $\sqrt{3}\pi r^3 \div 8r^3 = \sqrt{3}\pi/8$ or about 0.680.... So this arrangement which is known as the body-centred cubic lattice has a packing of 68%.

A further possibility is where there are additional atoms in the centres of the faces of the cube rather than at the centre. This gives the face-centred cubic lattice. For equal touching spheres their radius now has to be $1/2\sqrt{2}r$ so that their volume is $1/3\sqrt{2}\pi r^3$ each. Each cell now contains one-eighth of each of eight

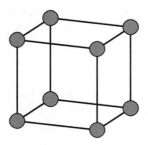

Simple cubic lattice, packing is 52%.

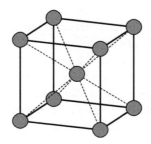

Body-centred cubic lattice, packing is 68%.

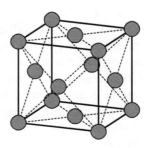

Face-centred cubic lattice, packing is 74%.

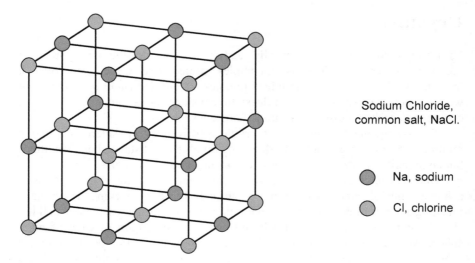

Sodium Chloride, common salt, NaCl.

● Na, sodium

○ Cl, chlorine

spheres plus one-half of each of six spheres – total four spheres and so the total volume is $^4/_3\sqrt{2}\pi r^3$. It follows that the proportion occupied by the spheres is $^4/_3\sqrt{2}\pi r^3 \div 8r^3 = \sqrt{2}\pi/6$ or about 0.740.... Gosh, that is a familiar number and indeed by looking at this packing carefully we find that it is the same arrangement as discussed earlier. The planes containing the faces of the cubes exhibit the square tiling (but at 45° to the edges of the cubes) and the planes containing the faces of the inscribed tetrahedra (these planes are perpendicular to the space diagonals) exhibit the triangular tiling.

Common salt, Sodium Chloride, NaCl, has the sodium and chlorine atoms at alternate vertices of the simple cubic lattice. In the diagram above the sodium atoms are shown in red and the chlorine atoms are in green. Another way of looking at this structure is to observe that the chlorine atoms occur at the points of a face-centred cubic lattice. So it comprises two such face-centred lattices interleaved.

Salt normally forms small cubic crystals. However, although the lattice is primarily cubic, crystallization can occur along other planes. The preferred planes are those with a high density of atoms. With a unit side to a cube, then each unit area of a plane containing the faces of the cubes exhibits the square tiling and contains 1 atom per unit area. However, if we take a plane defined by opposite edges of a cube, then we get a tiling of rectangles whose edges are 1 and √2. On such a plane the density is 1/√2 or about 0.707 atoms per unit area. Another possibility is to take a plane defined by one of the faces of an inscribed tetrahedron (this of course is the same direction as the faces of the dual octahedron). Such a plane exhibits atoms arranged as in the triangular tiling with edge √2. The density is 1/√3 or about 0.577 atoms per unit area. The interesting thing about these planes is that they only contain atoms of one kind.

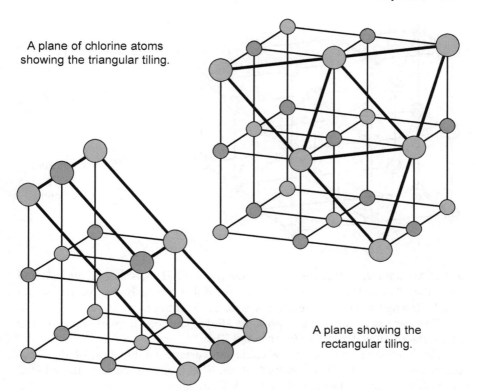

A plane of chlorine atoms showing the triangular tiling.

A plane showing the rectangular tiling.

The diagrams above show examples of both these planes; their atoms are shown larger for clarity. Note that under some circumstances salt does form octahedral crystals.

Many compounds such as Magnesium Oxide, MgO, show the same structure as Sodium Chloride. A related example is Iron Pyrites, FeS_2, which forms beautiful cubic crystals often called Fool's Gold. In this case the two Sulphur atoms are grouped together as one item at one set of locations whereas the Iron atoms are at the other set of locations.

The cubic structure is also shown by crystals formed of pure elements. Many metals such as Gold (Au), Silver (Ag) and Copper (Cu) have the close-packed face-centred cubic structure. Some metals such as Chromium (Cr) and Tungsten (W) have the looser body-centred cubic structure. The inert gases (as solids) such as Neon, Argon and Krypton take the face-centred cubic structure but the lighter Helium has the hexagonal close-packed structure mentioned when we were discussing the packing of spheres.

The body-centred cubic structure is adopted by a number of compounds and the classic model is Caesium Chloride, CsCl. The corners of the cubic cells are one atom such as Chlorine and the central atom is the other, Caesium. Another compound with this structure is Ammonium Chloride, NH_4Cl, commonly known

296 Gems of Geometry

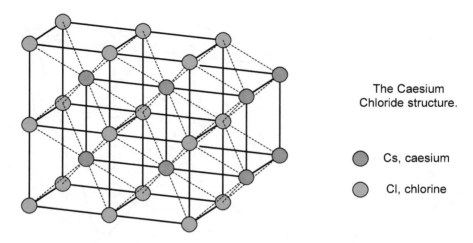

The Caesium Chloride structure.

● Cs, caesium

○ Cl, chlorine

as Sal Ammoniac. In this case the ammonium ion NH_4 acts as a single unit and takes the place of Caesium. It should be noted that this body-centred lattice can also be thought of as two interlaced simple cubic lattices.

The same structure occurs with some alloys. Thus one form of brass consists of interlaced cubic lattices of Copper and Zinc.

An interesting variant of the cubic lattice is a combination of the face-centred lattice and the body-centred lattice. This is the structure of the mineral Perovskite, Calcium Titanate, $CaTiO_3$. A Titanium atom is at the centre of the body, Oxygen atoms are in the centres of the faces and Calcium atoms are at the vertices of the cubic cell.

The Oxygen atoms are thus at the vertices of linked octahedra. It might look as if there were many more Calcium atoms than Titanium atoms but they occur

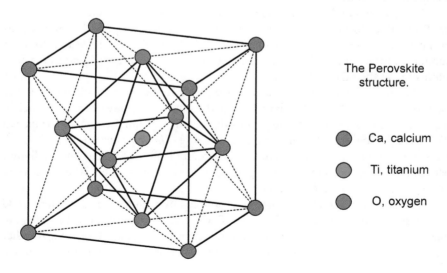

The Perovskite structure.

● Ca, calcium

● Ti, titanium

● O, oxygen

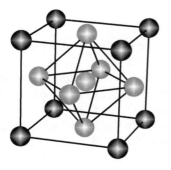 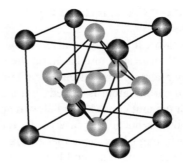

Calcium Titanate in stereo.

in equal numbers, one per cell. Remember that each atom at a corner of a cell is shared among eight cells. But there are three Oxygen atoms per cell since each atom is only shared among two cells. The arrangement is thus consistent with the formula $CaTiO_3$. Potassium Iodate, KIO_3, adopts the same structure.

A related structure is that of Fluorite, Calcium Fluoride, CaF_2. The Calcium atoms are arranged as a face-centred lattice and the Fluorine atoms are at the centres of the small cubes which subdivide the unit cube into eight.

Each Fluorine atom is linked to four Calcium atoms and each Calcium atom is linked to eight Fluorine atoms. Each unit cell contains eight Fluorine atoms one at the centre of each small cube. Each unit cell contains just four Calcium atoms. Again we have to remember that atoms on the face or at the corner of the unit cell are shared with adjacent cells. So the eight cells at the corners count as only one atom and the six cells on the faces count as three giving a total of four Calcium atoms which is consistent with the formula CaF_2.

Note how the Fluorine atoms together with the Calcium atoms at the centres of the faces of the cube form a rhombic dodecahedron.

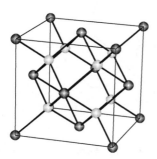 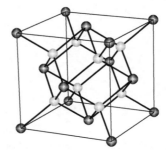

Calcium Fluoride in stereo.

Diamonds and graphite

DIAMOND can be looked at in many ways. In one way it is closely related to the fluorite structure based on the face-centred lattice. The key difference is that only alternate smaller cubes contain an atom. This structure is shown below. Each red carbon atom is linked with a covalent bond to four blue ones and vice versa. In the unit cube shown below it appears that only the blue atoms are linked to four others but in the case of the red atoms they also link to atoms in other cubes. Although it is clear that the red atoms form a face-centred lattice, it is perhaps surprising that the blue ones also form a face-centred lattice interlaced with the red one.

Perhaps a better way of looking at the diamond structure is as touching tetrahedra as shown opposite. One carbon atom (in blue) is at the centre of each tetrahedron and its four neighbours are in red. Note that there are also interlinked slightly buckled hexagonal tilings – alternate atoms in each ring are blue and red. There are four such tilings corresponding to the four faces of a tetrahedron.

Equally, we can look upon the structure as a series of tetrahedra with the apex down and with a red atom at the centre. In either case the tetrahedra are arranged as those of one orientation in the honeycomb of tetrahedra and octahedra.

It is perhaps surprising that the tetrahedral structure is the same as that of the interlaced face-centred lattices. However, recall that a tetrahedron can be circumscribed by a cube. If we do that for each tetrahedron of red atoms then the cubes form a honeycomb with alternate cubes missing and that is precisely the form of one face-centred lattice. The blue atoms at the centres of the tetrahedra form the other face-centred lattice.

This is much easier to understand by looking at the stereo image shown opposite and below. This comprises two complete face-centred cubes interlaced,

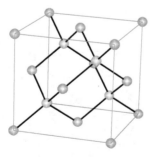 

The diamond structure is similar to fluorite with half the fluorine atoms omitted.

The covalent bonds within the unit cube are shown in black. The red lines merely denote the boundary of the unit cube.

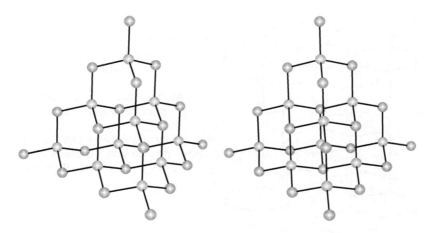

The structure of diamond looked at as touching tetrahedra in stereo. A carbon atom is at the vertices and at the centre of each tetrahedron.

one of red atoms and one of blue atoms. In order to help visualization, the outlines of parts of the cubic cells are shown as fine red and blue lines. Note that the cubes are shown at a slightly different orientation than usual – there is a somewhat greater rotation about the vertical axis in order to avoid the front atoms obscuring some of those behind. This stereo image shows the buckled hexagonal rings quite clearly.

Incidentally, the diamond structure is not densely packed. It is easily shown that the density is $\pi\sqrt{3}/16$ which is about 34%.

The elements Silicon, Germanium, and Tin (white tin) also adopt the diamond structure. Some compounds such as the mineral Zinc Blende (Zinc Sulphide), ZnS, have the same structure with each atom of Zinc connected to four atoms of Sulphur and vice versa.

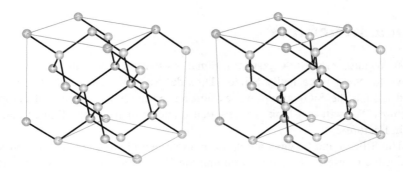

The diamond structure as two interlinked face-centred lattices in stereo.

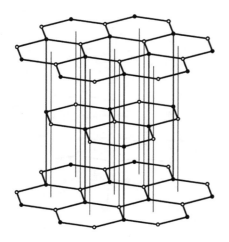

Graphite consists of layers of hexagonal tilings of carbon atoms.

Atoms in black are aligned with atoms in adjacent layers.

Atoms with white centres are aligned with the centres of hexagons in adjacent layers.

Graphite is another well-known form of carbon. The atoms are arranged as layers of hexagonal tilings. The layers are more than twice as far apart as the individual atoms within a layer but are shown closer together above to save space. The bonding between layers is weak which is why graphite is an excellent lubricant. Moreover, there are two types of atoms which are not structurally equivalent. Some, shown in black above, are directly aligned with atoms in adjacent layers. Others, shown with white centres, are aligned with the centres of the hexagonal rings in adjacent layers.

A third allotrope of carbon was recently discovered as mentioned in Lecture 2. The carbon atoms are arranged as the vertices of a truncated icosahedron to form the molecule C_{60}. Related structures are nanotubes; the basic form is obtained by breaking a C_{60} molecule into two hemispheres and then joining them by a tube of further carbon atoms arranged as a rolled up sheet of hexagons like a tube of graphite one atom thick. Such tubes are immensely strong.

Silica structures

THE WALRUS AND THE CARPENTER remarked that there was much sand in the world. Sand is basically Silicon Dioxide, SiO_2 or quartz. Most rocks are based around the SiO_2 group and exhibit an extraordinary variety of complex structures. We will just look at a few that relate to many of the figures that we met in Lecture 2.

The Silicon and Oxygen atoms form a tetrahedral structure with the Silicon atom at the centre of the tetrahedron and the Oxygen atoms at the four vertices. Many such tetrahedra can then be linked together by their vertices. Each Oxygen atom is then shared by two Silicon atoms so that the formula SiO_2 is satisfied.

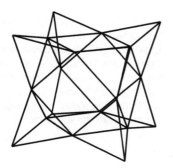 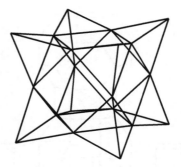

Eight tetrahedra linked together as if around a cuboctahedron in a Zeolite.

In one form of Zeolite, eight such tetrahedra are linked together in a cubic arrangement as if they were on the faces of a cuboctahedron as shown above. These octuplet groups are then linked together by their free vertices to form the amazing structure shown below. Each unit cube contains a truncated cuboctahedron and each vertex is the location of one tetrahedron. Six of the eight tetrahedra forming an octuplet are emphasized as black circles. The other two are on the surface where the two unit cubes in front of those shown are located.

The tetrahedra thus form rings of four, six and eight. The structure has large holes in it and is known as a molecular sieve. Simple molecules/ions such as water can flow through the sieve but larger ones become trapped. This is the basis of chemical water softeners. Large groups containing Calcium become trapped and thus filtered from the water. When the material is sufficiently clogged, Sodium Chloride is introduced and the Calcium and Sodium ions are interchanged. The smaller sodium based group can then be flushed out as well as the resulting Calcium Chloride. Incidentally, the structure is a honeycomb of truncated cuboctahedra, truncated octahedra and cubes. This is somewhat irregular and was not discussed in Lecture 2 since we required all vertices and lines of a honeycomb to have the same arrangement of polyhedra about them.

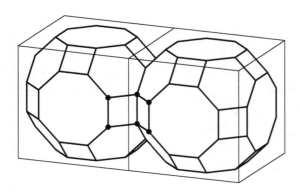

Two unit cubes containing truncated cuboctahedra illustrate the structure of a molecular sieve.

302   Gems of Geometry

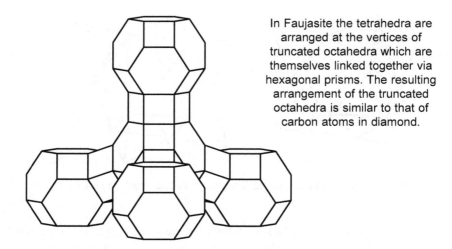

In Faujasite the tetrahedra are arranged at the vertices of truncated octahedra which are themselves linked together via hexagonal prisms. The resulting arrangement of the truncated octahedra is similar to that of carbon atoms in diamond.

In Faujasite, some tetrahedral cells have the Silicon atoms replaced by Aluminium. The cells are linked together via the common Oxygen atoms to form truncated octahedra. These in turn are then linked together via hexagonal prisms in a large tetrahedral form as shown above. The overall superstructure is exactly as diamond. Again the structure is very open.

We conclude this brief discussion of silica structures by considering silica itself. This exists in various forms and perhaps the most interesting is β-quartz. It is again composed of linked tetrahedral cells consisting of a Silicon atom at the centre and Oxygen atoms at each vertex. However, it is more revealing to consider the individual atoms rather than the tetrahedra.

The structure is essentially a number of spirals with three Silicon atoms per rotation. If we imagine the spirals as being vertical then the plan view is as shown below. Note that the Silicon atoms (in black) in this plan view are arranged as the tiling {3.6.3.6} described in Lecture 2. Each spiral has three adjacent spirals and they are attached to each other via common Silicon atoms.

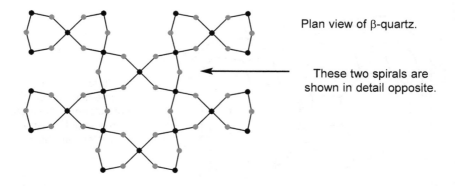

Plan view of β-quartz.

These two spirals are shown in detail opposite.

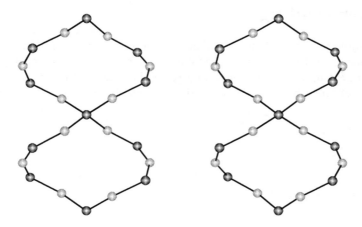

Two spirals of quartz in stereo.

Note that six spirals surround a hexagonal space; it is thus not surprising that quartz crystallizes as hexagonal prisms.

The diagram above shows two adjacent spirals in stereo. They twist in a clockwise manner and two complete rotations are shown. The adjacent spirals share a Silicon atom once per rotation. Quartz can equally crystallize with anticlockwise spirals and so exhibits enantiomorphism. Sometimes crystals occur as twins with a part of each form.

Gems

WE CONCLUDE with a few remarks about some of the more well known gemstones. Ruby and Sapphire are simply the clear mineral Corundum, Al_2O_3, with traces of other elements which impart the colour. The Oxygen atoms are arranged as the hexagonal close-packed structure with the Aluminium atoms squeezed in some of the gaps between the larger Oxygen atoms.

Emerald is Beryl with traces of Vanadium or Chromium imparting the green colour. Beryl is complex with formula $Be_3Al_2Si_6O_{18}$. The Silicon and Oxygen atoms are arranged in rings in the shape of a puckered hexagon – these rings form the familiar tiling of hexagons in a series of parallel layers. The Beryllium and Aluminium atoms connect to the Oxygen atoms in adjacent layers and thereby connect the layers together.

Amethyst is simply Quartz with traces of Iron giving the colour. Fluorspar (Calcium Fluoride, see above) normally occurs as cubic crystals but examples of twin cubes arranged as in the compound of five cubes discussed in Lecture 2 have been found.

Further reading

THE PROOF of the packing of spheres is described at length in *Kepler's Conjecture* by George Szpiro. There are so many books on crystallography that it is rather hard to recommend any particular one for further reading but volume IV of *The Crystalline State* by Sir Lawrence Bragg is a classic text.

F Stability

THIS APPENDIX looks in more detail at the stability of various aspects of the system discussed in Lecture 8.

Stability of fixed points

WE START by considering the stability of fixed points of a function $y = f(x)$. The stability of a fixed point is determined by the derivative at that point. Remember that the derivative is the slope or gradient usually written as df/dx and often abbreviated to $f'(x)$.

The rules regarding stability are

$f' > 1$	unstable, diverges monotonically
$0 < f' < 1$	stable, converges monotonically
$-1 < f' < 0$	stable, oscillates and converges
$f' < -1$	unstable, oscillates and diverges

So when f' is positive, the behaviour is monotonic and when f' is negative the system oscillates. Moreover, if the absolute value of f' is less than 1 then it converges. Note that when the gradient is +1 or −1 a transition occurs between stability and instability. These are in fact typically the bifurcation points.

The reason for the behaviour at a fixed point is easily understood by considering a diagram such as below. The points where the function $f(x)$ meets the line $y = x$ are the fixed points. This is obvious because the function is defined by $y = f(x)$ and a fixed point is one where $f(x) = x$.

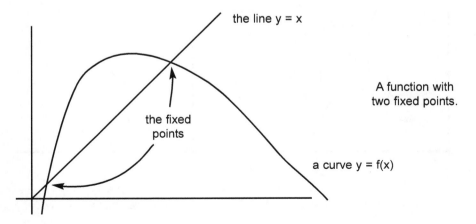

A function with two fixed points.

306 Gems of Geometry

We now consider the four cases separately and for simplicity we will just look at a small part of where the function f crosses the line $y = x$. In fact we can take a small portion of f to be a straight line. Now suppose we start with an initial value x_0 which is slightly greater than the fixed point. Applying the function f we get the next point $x_1 = f(x_0)$. This can be determined graphically by simply drawing a vertical line from x_0 on the x-axis up to the curve and then a horizontal line to meet the line $y = x$. This gives the next value, x_1. We then repeat this process to give x_2, x_3 and so on.

The diagrams below show the four cases with $f'(x)$ equal to -0.5, $+0.5$, -2, and $+2$. The reason for the different behaviours should now be clear.

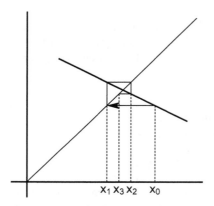

f' = –1/2,
oscillates and converges.

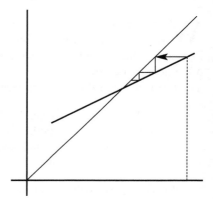

f' = +1/2,
converges monotonically.

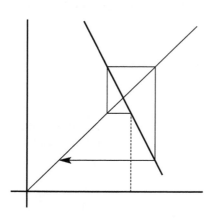

f' = –2,
oscillates and diverges.

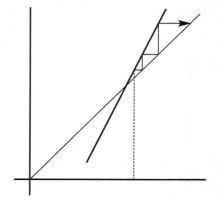

f' = +2,
diverges monotonically.

The fixed points

WE WILL NOW consider the population function studied in Lecture 8 whose behaviour depends upon the value of the parameter μ, namely

$$f(x) = \mu x(1 - x)$$

We consider various positive values of μ.
There are two fixed points given by

$f(x) = x$ or in other words $x = \mu x(1 - x)$ which is

$$\mu x^2 + (1 - \mu)x = 0$$

This is a quadratic equation and so has two roots. One is at $x = 0$ and the other is at $x = (\mu - 1)/\mu$ or $1 - 1/\mu$.

As we have seen, the stability of a fixed point is determined by the derivative at that point. Recall that the derivative of x^n is $n \times x^{n-1}$. Thus the derivative of μx is just μ and the derivative of μx^2 is $2\mu x$. So we have

$$f'(x) = \mu - 2\mu x = \mu(1 - 2x)$$

Hence the derivative of the fixed point at $x = 0$ is $\mu(1 - 2 \times 0) = \mu$. For μ between 0 and 1, this is clearly less than 1 and so is stable. For μ above 1, it is unstable and diverges monotonically.

For the fixed point at $x = 1 - 1/\mu$, the derivative is

$$\mu(1 - 2(1 - 1/\mu)) = \mu(2/\mu - 1) = 2 - \mu$$

Note that if $\mu < 1$, the corresponding value of x is negative and so of no interest. For $1 < \mu < 2$ the derivative is positive and less than 1. It is a stable point and converges monotonically. Here is the behaviour starting at 0.1 for the case $\mu = 1.5$ where the fixed point is at $1 - 1/\mu = 1/3$.

$\mu = 1.50000$

```
0.10000  0.13500  0.17516  0.21672
0.25463  0.28469  0.30546  0.31823
0.32544  0.32929  0.33129  0.33230
0.33282  0.33308  0.33320  0.33327
0.33330  0.33332  0.33333  0.33333
0.33333  0.33333  0.33333  0.33333
```

At $\mu = 2$ the derivative is zero and then changes sign. For $2 < \mu < 3$, the derivative lies between -1 and 0. It is still stable but converges in an oscillatory manner. Thus

$\mu = 2.50000$

0.10000 0.22500 0.43594 0.61474
0.59209 0.60380 0.59806 0.60096
0.59952 0.60024 0.59988 0.60006
0.59997 0.60002 0.59999 0.60000
0.60000 0.60000 0.60000 0.60000

Note how it overshoots but then rapidly settles down.

When μ is exactly 3, the derivative is -1. This is the first bifurcation point and convergence is extremely slow. The first 50 values are

$\mu = 3.000$

0.100 0.270 0.591 0.725 0.598 0.721 0.603 0.718 0.607 0.715
0.611 0.713 0.614 0.711 0.616 0.709 0.618 0.708 0.620 0.707
0.622 0.705 0.623 0.704 0.625 0.703 0.626 0.702 0.627 0.702
0.628 0.701 0.629 0.700 0.630 0.699 0.631 0.699 0.632 0.698
0.632 0.697 0.633 0.697 0.634 0.696 0.634 0.696 0.635 0.695

and even after 100,000 iterations it has still not converged to three decimal digits, the last few being

0.665920 0.667411 0.665920 0.667411 0.665921
0.667411 0.665921 0.667411 0.665921 0.667411
0.665921 0.667411 0.665921 0.667411 0.665921
0.667411 0.665921 0.667411 0.665921 0.667411

It does eventually converge to 2/3 of course.

It is quite easy to investigate just why convergence is so slow in this instance. Suppose we start at $x = 2/3 + \delta$ where δ is small. Then the next point is given by

$$f(x) = 3x(1 - x)$$
$$= (2 + 3\delta)(1/3 - \delta)$$
$$= 2/3 - \delta - 3\delta^2$$

This is the other side to where we started and there is also the term $3\delta^2$. In fact if δ is positive then we are actually further away. So let's take two iterations so that we end up on the same side.

$$f(x) = 3x(1 - x)$$
$$= (2 - 3\delta - 9\delta^2)(1/3 + \delta + 3\delta^2)$$
$$= 2/3 + \delta - 18\delta^3 - 27\delta^4$$

Note that the terms in δ^2 cancel. If δ is small then we can ignore the term in δ^4 so after every two iterations we are $18\delta^3$ nearer. Suppose for example that δ is 0.001 (which is about the position after 100,000 iterations). Then two more iterations will bring us 0.000000018 nearer. It's not surprising that convergence is slow. In fact the closer we get, the slower the convergence becomes.

We can compare this with convergence around $\mu = 2$. The fixed point is 1/2 so we start at $1/2 + \delta$. We get the next point

$$f(x) = 2x(1 - x)$$
$$= (1 + 2\delta)(1/2 - \delta)$$
$$= 1/2 - 2\delta^2$$

In this case the terms in δ cancel. So if we start with $\delta = 0.001$ then the error at the next iteration is only 0.000002. So the number of exact decimal places nearly doubles each time. So the closer we get, the faster the convergence becomes.

When $\mu > 3$ the fixed point becomes unstable. It is still there of course and if we choose numbers that can be represented exactly in a binary computer then we can demonstrate it. Thus taking $\mu = 4$ the fixed point is at $x = 0.75$. Both these numbers can be held exactly and the computation of $4.0x(1-x)$ can also be done exactly.

Starting at exactly 0.75 we get

$\mu = 4.0000000$

0.7500000 0.7500000 0.7500000 0.7500000
0.7500000 0.7500000 0.7500000 0.7500000
0.7500000 0.7500000 0.7500000 0.7500000
0.7500000 0.7500000 0.7500000 0.7500000

But starting at 0.7500001 we get

$\mu = 4.0000000$

0.7500001 0.7499998 0.7500005 0.7499990
0.7500019 0.7499962 0.7500076 0.7499847
0.7500305 0.7499390 0.7501221 0.7497558
0.7504882 0.7490227 0.7519507 0.7460833
0.7577720 0.7342143 0.7805746 0.6851115
0.8629349 0.4731131 0.9971083 0.0115332
0.0456007 0.1740852 0.5751181 0.9774291
0.0882458 0.3218341 0.8730276 0.4434018
0.9871866 0.0505971 0.1921480 0.6209086
0.9415245 0.2202244 0.6869025 0.8602698

Here we see that it diverges by oscillating around 0.75. Chaos soon sets in.

The two cycle

BIFURCATION SETS IN at $\mu = 3$. The fixed point becomes unstable and is effectively replaced by a stable two cycle.

The two cycle is defined by the function $f(f(x))$ which is often written as $f^{(2)}(x)$.

$$f^{(2)}(x) = f(f(x)) = \mu f(x)(1 - f(x)) = \mu^2 x(1 - x)(1 - \mu x(1 - x))$$
$$= \mu^2(x - x^2)(1 - \mu x + \mu x^2) = -\mu^3 x^4 + 2\mu^3 x^3 - (\mu^3 + \mu^2)x^2 + \mu^2 x$$

For a point x to be on a two cycle we must have $f(f(x)) = x$, so we get

$$x = -\mu^3 x^4 + 2\mu^3 x^3 - (\mu^3 + \mu^2)x^2 + \mu^2 x \quad \text{which rearranges to}$$
$$\mu^3 x^4 - 2\mu^3 x^3 + (\mu^3 + \mu^2)x^2 - (\mu^2 - 1)x = 0$$

One root of this is $x = 0$. Note that any fixed point is also a two cycle because two applications of the function clearly leave the value unchanged. It therefore follows that the other fixed point is also a root of the equation and indeed it can be factorized to give

$$x(\mu x - \mu + 1)[\mu^2 x^2 - \mu(\mu + 1)x + (\mu + 1)] = 0$$

So the quadratic equation

$$\mu^2 x^2 - \mu(\mu + 1)x + (\mu + 1) = 0$$

corresponds to the true two cycle. It will be recalled that the quadratic equation $ax^2 + bx + c = 0$ has real roots only if the so-called discriminant $b^2 - 4ac$ is zero or positive. In this case the discriminant is

$$\mu^2(\mu + 1)^2 - 4\mu^2(\mu + 1) = \mu^2(\mu + 1)(\mu - 3)$$

This is positive only if $\mu > 3$ which shows that genuine period two points only exist if $\mu > 3$. (If $\mu = 3$ we get a double root.) We ignore negative values of μ.

Solving the quadratic equation in the usual way we find that the roots are

$$x = \frac{\mu(\mu+1) \pm \sqrt{[\mu^2(\mu+1)^2 - 4\mu^2(\mu+1)]}}{2\mu^2}$$
$$= \frac{(\mu+1) \pm \sqrt{[(\mu+1)(\mu-3)]}}{2\mu}$$

and so these are the points of the two cycle.

The stability of the two cycle can be investigated by considering the derivative of $f(f(x))$. This derivative is

$$-4\mu^3 x^3 + 6\mu^3 x^2 - 2(\mu^3 + \mu^2)x + \mu^2$$

Remember that the two cycle starts when $\mu = 3$ at which point the two roots of the quadratic equation are equal and in fact both are 2/3 which is also the fixed point which is about to go unstable. Indeed the equation $f(f(x)) = x$ actually has a triple root of 2/3.

At this point ($\mu = 3$, $x = 2/3$) the derivative of the two cycle is

$$-4\times 3^3 \times (2/3)^3 + 6\times 3^3 \times (2/3)^2 - 2(3^3 + 3^2)(2/3) + 3^2 = -32 + 72 - 48 + 9 = 1$$

So the derivative is +1 which is characteristic of a bifurcation point. The derivative is −1 at the next bifurcation point which is where the two cycle becomes unstable and breaks into a four cycle. At this bifurcation point we therefore have

$$-4\mu^3 x^3 + 6\mu^3 x^2 - 2(\mu^3 + \mu^2)x + \mu^2 = -1 \qquad f^{(2)\prime}(x) = -1$$

and of course we also have

$$\mu^2 x^2 - \mu(\mu + 1)x + (\mu + 1) = 0 \qquad f^{(2)}(x) = x$$

So we now have two equations for μ and x. These look a bit tricky but are in fact quite easy. It helps to put $y = \mu x$ in order to reduce the number of instances of μ. We then get

$$-4y^3 + 6\mu y^2 - 2(\mu^2 + \mu)y + \mu^2 = -1 \qquad (1)$$

$$y^2 - (\mu + 1)y + (\mu + 1) = 0 \qquad (2)$$

Now multiply equation (2) by $4y$ and add it to equation (1). This eliminates the terms in y^3 and gives

$$(2\mu - 4)y^2 + (-2\mu^2 + 2\mu + 4)y + \mu^2 + 1 = 0 \qquad (3)$$

Now multiply equation (2) by $(2\mu - 4)$ and subtract it from equation (3). Rather surprisingly all the terms in y now vanish and we are simply left with

$$\mu^2 - 2\mu - 5 = 0$$

and so

$$\mu = 1 + \sqrt{6} = 3.44948974...$$

Indeed we can easily check that 3.44 is still a two cycle whereas 3.45 is a four cycle. Thus

$\mu = 3.44000$

0.10000 0.30960 0.73529 0.66955
0.76111 0.62547 0.80584 0.53822

```
0.85497  0.42654  0.84143  0.45897
0.85421  0.42840  0.84237  0.45678
```

...

```
0.84851  0.44219  0.84850  0.44219
0.84851  0.44219  0.84850  0.44219
0.84851  0.44219  0.84850  0.44219
0.84851  0.44219  0.84850  0.44219
0.84851  0.44219  0.84850  0.44219
```

and

$\mu = 3.45000$

```
0.10000  0.31050  0.73861  0.66608
0.76734  0.61592  0.81614  0.51768
0.86142  0.41184  0.83569  0.47373
0.86012  0.41508  0.83762  0.46924
```

...

```
0.85244  0.43397  0.84746  0.44599
0.85244  0.43397  0.84746  0.44599
0.85244  0.43397  0.84746  0.44599
0.85244  0.43397  0.84746  0.44599
0.85244  0.43397  0.84746  0.44599
```

This converges quite slowly to its four cycle because 3.45 is close to the bifurcation point.

As μ increases, bifurcation continues at an ever increasing rate giving an 8-cycle at 3.55, a 16-cycle at 3.566, a 32-cycle at 3.569, a 64-cycle at 3.5696 and so on.

The bifurcation points can be tabulated thus

1-2	3.00000
2-4	3.44949
4-8	3.54408
8-16	3.56438
16-32	3.56867

It is interesting to compute the ratio of the differences between these successive pairs of values. In fact they converge to a constant which is about 4.669. This curious constant is known as the Feigenbaum number δ after the American theoretical physicist, Mitchell Feigenbaum as was mentioned in Lecture 8.

Three cycles

WE CAN INVESTIGATE the three cycle in a similar manner although it gets rather tedious. We have to solve

$$f(f(f(x))) = x \quad \text{or} \quad f^{(3)}(x) = x$$

We get an eighth degree equation and work in stages, thus

$$f(f(f(x))) = \mu f^{(2)}(x)(1 - f^{(2)}(x)) = x$$

which eventually becomes

$$\mu^7 x^8 - 4\mu^7 x^7 + (6\mu^7 + 2\mu^6)x^6 - (4\mu^7 + 6\mu^6)x^5 + (\mu^7 + 6\mu^6 + \mu^5 + \mu^4)x^4 \\ - (2\mu^6 + 2\mu^5 + 2\mu^4)x^3 + (\mu^5 + \mu^4 + \mu^3)x^2 - (\mu^3 - 1)x = 0$$

As before we know that $x = 0$ and $\mu x = \mu - 1$ are roots because a fixed point is also a three cycle. We can therefore remove these factors. And again we put $y = \mu x$ in order to reduce the number of instances of μ.

We finally get

$$y^6 - (3\mu + 1)y^5 + (3\mu + 1)(\mu + 1)y^4 - (\mu^3 + 5\mu^2 + 3\mu + 1)y^3 + \\ (2\mu + 1)(\mu^2 + \mu + 1)y^2 - (\mu^2 + \mu + 1)(\mu + 1)y + (\mu^2 + \mu + 1) = 0$$

This is a sixth degree equation and so has six roots. We know that if x is a root then $f(x)$ and $f(f(x))$ are also roots. It follows therefore that the six roots form two three cycles. Clearly, if one root of a cycle is real then the other two are real as well. Moreover, since complex roots come in pairs it follows that if one cycle is real then at least four roots must be real and hence that the other cycle is also real. So either all the roots are real or all the roots are complex (and in conjugate pairs).

If we plot the locus of the roots in the complex plane as μ varies we will obtain a pattern much like that shown below.

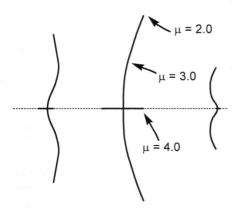

A typical root locus diagram. The six roots move as μ varies. The complex roots are in conjugate pairs and so the complex branches are symmetric. The branches hit the real axis when $\mu = 2\sqrt{2}+1$ and then the pairs of real roots move along the axis in opposite directions. This shows the movement from $\mu = 2.0$ to $\mu = 4.0$.

The three pairs of branches hit the x-axis together at the value of μ when the six complex roots become real. Clearly at that value of μ the roots are double. So the equation must take the form

$$(y^3 + Ay^2 + By + C)^2 = 0 \text{ or}$$
$$y^6 + 2Ay^5 + (A^2 + 2B)y^4 + (2C + 2AB)y^3 + (2AC + B^2)y^2 + 2BCy + C^2 = 0$$

We can now compare the coefficients of the various powers of y and obtain

$$2A = -(3\mu + 1) \quad (1)$$
$$A^2 + 2B = (3\mu + 1)(\mu + 1) \quad (2)$$
$$2C + 2AB = -(\mu^3 + 5\mu^2 + 3\mu + 1) \quad (3)$$
$$2AC + B^2 = (2\mu + 1)(\mu^2 + \mu + 1) \quad (4)$$
$$2CB = -(\mu^2 + \mu + 1)(\mu + 1) \quad (5)$$
$$C^2 = \mu^2 + \mu + 1 \quad (6)$$

So here we have six equations for the four unknowns A, B, C and μ. Clearly there is some redundancy.

Eliminating A between equations (1) and (2) we get

$$8B = (3\mu + 1)(\mu + 3) \quad (7)$$

and eliminating C between equations (5) and (6) we get

$$4B^2 = (\mu^2 + \mu + 1)(\mu + 1)^2 \quad (8)$$

We can now eliminate B between equations (7) and (8) and finally obtain an equation for μ

$$(3\mu + 1)^2(\mu + 3)^2 - 16(\mu^2 + \mu + 1)(\mu + 1)^2 = 0$$

This is a quartic equation and after multiplying out the terms it becomes

$$7\mu^4 - 12\mu^3 - 54\mu^2 - 12\mu + 7 = 0$$

Note the symmetry in this equation. It follows that if m is one root then $1/m$ is another root. This enables us to deduce that it can be factorized thus

$$(\mu^2 - 2\mu - 7)(7\mu^2 + 2\mu - 1) = 0$$

Finally, we find the roots to be

$$2\sqrt{2} + 1, -2\sqrt{2} + 1, (2\sqrt{2} - 1)/7, (-2\sqrt{2} - 1)/7$$

As is usual in these situations only one of these is significant. In fact it is the first one. So the value of μ when the three cycle first appears is $2\sqrt{2} + 1$ or about 3.828427....

Just below this there is chaos and just above it there is a stable three cycle. At the critical value there is in fact a double three cycle. As μ increases this double three cycle separates into two distinct three cycles quite close together. Only one of these is stable.

We can actually find expressions for the values of the three cycle. First, using the known value of μ after some manipulation we have

$$A = -3\sqrt{2} - 2$$
$$B = 5 + 4\sqrt{2}$$
$$C = -3 - \sqrt{2}$$

and of course we have

$$y^3 + Ay^2 + By + C = 0$$

We now make the usual substitution $z = y + A/3$ in order to remove the quadratic term. After much manipulation we obtain

$$z^3 - 7z/3 - 7/27 = 0$$

We use the normal technique for solving such cubics by putting $z = u + v$ where uv equals one-third of the coefficient of z; so $uv = -7/9$ and we eventually find

$$z = \sqrt[3]{(7/54)} \, [\sqrt[3]{(1 + 3\sqrt{3}i)} + \sqrt[3]{(1 - 3\sqrt{3}i)}]$$

This is the awkward case where although the roots are real they turn out as expressions involving the cube root of complex numbers.

Undaunted we can seek the cube root of the complex number $1 + 3\sqrt{3}i$ geometrically. The real part is 1 and the imaginary part is $\sqrt{27}$. Using Pythagoras we thus find that the argument θ is given by $\tan \theta = \sqrt{27}$ and the modulus is $\sqrt{28}$.

The cube root is obtained by dividing the argument by 3 and taking the cube root of the modulus. We note moreover that we have to add the cube root of the conjugate number (that is the one with a minus sign for the imaginary part). When we add these two cube roots, the imaginary parts cancel and we are left simply with double the real part.

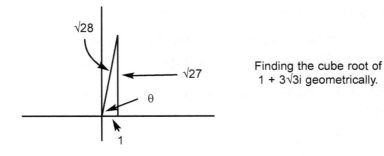

Finding the cube root of $1 + 3\sqrt{3}i$ geometrically.

This is

$$2 \times \sqrt[6]{28} \cos \theta/3 \quad \text{where } \tan \theta = \sqrt{27}$$

Then we have to multiply by the term $\sqrt[3]{(7/54)}$. This actually improves things since the factor becomes

$$2 \times \sqrt[6]{28} \times \sqrt[3]{(7/54)}$$
$$= 2 \times \sqrt[6]{((28 \times 49)/(54 \times 54))}$$
$$= 2\sqrt{7/3}$$

So finally we have a value for z namely

$$z = 2\sqrt{7/3} \cos \theta/3 \quad \text{where } \tan \theta = \sqrt{27}$$

There are of course two other values for z corresponding to the other points of the cycle and these are found by taking the other cube roots which are simply

$$z = 2\sqrt{7/3} \cos (2\pi/3 + \theta/3)$$
$$z = 2\sqrt{7/3} \cos (4\pi/3 + \theta/3)$$

We then obtain $y = z - A/3 = z + \sqrt{2} + 2/3$ and finally $x = y/\mu$. The values of x are

0.159928818..., 0.514355277..., 0.956317842...

So these are the values of the three cycle when it first appears at $\mu = 2\sqrt{2} + 1$ which is about 3.828427....

If we take a value of μ just slightly above this critical value then the stable and unstable three cycles are approximately equally spaced around the above values. For example if we take $\mu = 3.8285$ then the unstable three cycle is about

0.16072485, 0.51643545, 0.95609083   unstable cycle

and the stable cycle is

0.15912637, 0.51227308, 0.95654832   stable cycle

The variations from the double three cycle are

0.00079603, 0.00208017, 0.00022701   and
0.00080245, 0.00208220, 0.00023048   respectively

Finally, here is the behaviour starting at the above unstable three cycle in a rather smaller font

$\mu = 3.82850000$

F Stability 317

0.16072485	0.51643545	0.95609083		0.16071152	0.51640083	0.95609518
0.16072485	0.51643545	0.95609083		0.16070965	0.51639596	0.95609579
0.16072485	0.51643545	0.95609083		0.16070751	0.51639041	0.95609649
0.16072485	0.51643545	0.95609083		0.16070508	0.51638410	0.95609728
0.16072485	0.51643545	0.95609083		0.16070232	0.51637691	0.95609818
0.16072485	0.51643544	0.95609083		0.16069917	0.51636873	0.95609921
0.16072485	0.51643544	0.95609083		0.16069559	0.51635942	0.95610038
0.16072485	0.51643544	0.95609083		0.16069151	0.51634884	0.95610170
0.16072485	0.51643544	0.95609083		0.16068689	0.51633682	0.95610320
0.16072485	0.51643544	0.95609083		0.16068163	0.51632317	0.95610491
0.16072484	0.51643544	0.95609083		0.16067567	0.51630769	0.95610685
0.16072484	0.51643543	0.95609083		0.16066892	0.51629013	0.95610904
0.16072484	0.51643543	0.95609083		0.16066127	0.51627025	0.95611152
0.16072484	0.51643543	0.95609083		0.16065261	0.51624776	0.95611432
0.16072484	0.51643542	0.95609083		0.16064283	0.51622235	0.95611747
0.16072484	0.51643542	0.95609083		0.16063180	0.51619369	0.95612103
0.16072484	0.51643541	0.95609083		0.16061938	0.51616141	0.95612503
0.16072483	0.51643541	0.95609084		0.16060541	0.51612511	0.95612952
0.16072483	0.51643540	0.95609084		0.16058974	0.51608439	0.95613454
0.16072483	0.51643539	0.95609084		0.16057220	0.51603880	0.95614014
0.16072483	0.51643538	0.95609084		0.16055262	0.51598792	0.95614638
0.16072482	0.51643537	0.95609084		0.16053083	0.51593127	0.95615331
0.16072482	0.51643536	0.95609084		0.16050665	0.51586842	0.95616096
0.16072481	0.51643535	0.95609084		0.16047993	0.51579895	0.95616938
0.16072481	0.51643534	0.95609084		0.16045051	0.51572247	0.95617861
0.16072480	0.51643532	0.95609085		0.16041827	0.51563864	0.95618867
0.16072479	0.51643530	0.95609085		0.16038311	0.51554722	0.95619959
0.16072478	0.51643528	0.95609085		0.16034499	0.51544807	0.95621136
0.16072478	0.51643525	0.95609085		0.16030389	0.51534118	0.95622396
0.16072476	0.51643523	0.95609086		0.16025987	0.51522668	0.95623736
0.16072475	0.51643519	0.95609086		0.16021306	0.51510490	0.95625150
0.16072474	0.51643516	0.95609087		0.16016366	0.51497636	0.95626630
0.16072472	0.51643512	0.95609087		0.16011194	0.51484177	0.95628166
0.16072470	0.51643507	0.95609088		0.16005826	0.51470207	0.95629747
0.16072468	0.51643501	0.95609089		0.16000305	0.51455835	0.95631357
0.16072466	0.51643495	0.95609089		0.15994680	0.51441189	0.95632981
0.16072463	0.51643488	0.95609090		0.15989004	0.51426409	0.95634604
0.16072460	0.51643480	0.95609091		0.15983335	0.51411643	0.95636208
0.16072456	0.51643471	0.95609092		0.15977728	0.51397039	0.95637778
0.16072452	0.51643460	0.95609094		0.15972241	0.51382743	0.95639300
0.16072448	0.51643448	0.95609095		0.15966924	0.51368889	0.95640759
0.16072442	0.51643434	0.95609097		0.15961824	0.51355596	0.95642146
0.16072436	0.51643418	0.95609099		0.15956978	0.51342967	0.95643451
0.16072429	0.51643400	0.95609101		0.15952418	0.51331079	0.95644668
0.16072422	0.51643380	0.95609104		0.15948165	0.51319990	0.95645793
0.16072412	0.51643357	0.95609107		0.15944231	0.51309733	0.95646826
0.16072402	0.51643330	0.95609110		0.15940621	0.51300320	0.95647767
0.16072390	0.51643299	0.95609114		0.15937334	0.51291746	0.95648617
0.16072377	0.51643264	0.95609118		0.15934360	0.51283989	0.95649382
0.16072362	0.51643225	0.95609123		0.15931686	0.51277015	0.95650066
0.16072344	0.51643179	0.95609129		0.15929296	0.51270779	0.95650674
0.16072324	0.51643127	0.95609136		0.15927170	0.51265233	0.95651213
0.16072301	0.51643068	0.95609143		0.15925288	0.51260322	0.95651688
0.16072275	0.51643001	0.95609152		0.15923628	0.51255991	0.95652105
0.16072246	0.51642923	0.95609161		0.15922169	0.51252185	0.95652470
0.16072212	0.51642835	0.95609172		0.15920892	0.51248852	0.95652789
0.16072173	0.51642735	0.95609185		0.15919776	0.51245941	0.95653068
0.16072129	0.51642620	0.95609199		0.15918804	0.51243404	0.95653309
0.16072079	0.51642490	0.95609216		0.15917959	0.51241199	0.95653519
0.16072021	0.51642340	0.95609235		0.15917226	0.51239285	0.95653701
0.16071956	0.51642170	0.95609256		0.15916591	0.51237627	0.95653858
0.16071881	0.51641976	0.95609280		0.15916041	0.51236193	0.95653994
0.16071796	0.51641754	0.95609308		0.15915566	0.51234954	0.95654111
0.16071698	0.51641501	0.95609340		0.15915156	0.51233884	0.95654212
0.16071587	0.51641213	0.95609376		0.15914803	0.51232961	0.95654299
0.16071461	0.51640884	0.95609418		0.15914498	0.51232167	0.95654374
0.16071317	0.51640510	0.95609465		0.15914236	0.51231482	0.95654439

```
0.15914010  0.51230893  0.95654494        0.15912642  0.51227322  0.95654831
0.15913816  0.51230387  0.95654542        0.15912641  0.51227320  0.95654831
0.15913649  0.51229951  0.95654583        0.15912641  0.51227319  0.95654831
0.15913506  0.51229577  0.95654618        0.15912640  0.51227317  0.95654831
0.15913383  0.51229255  0.95654649        0.15912640  0.51227316  0.95654831
0.15913277  0.51228979  0.95654675        0.15912639  0.51227315  0.95654831
0.15913186  0.51228742  0.95654697        0.15912639  0.51227314  0.95654831
0.15913108  0.51228538  0.95654716        0.15912639  0.51227313  0.95654831
0.15913041  0.51228363  0.95654733        0.15912638  0.51227312  0.95654831
0.15912984  0.51228213  0.95654747        0.15912638  0.51227312  0.95654832
0.15912934  0.51228084  0.95654759        0.15912638  0.51227311  0.95654832
0.15912892  0.51227974  0.95654769        0.15912638  0.51227311  0.95654832
0.15912856  0.51227879  0.95654778        0.15912638  0.51227310  0.95654832
0.15912824  0.51227798  0.95654786        0.15912638  0.51227310  0.95654832
0.15912798  0.51227728  0.95654792        0.15912637  0.51227310  0.95654832
0.15912775  0.51227668  0.95654798        0.15912637  0.51227310  0.95654832
0.15912755  0.51227617  0.95654803        0.15912637  0.51227309  0.95654832
0.15912738  0.51227573  0.95654807        0.15912637  0.51227309  0.95654832
0.15912724  0.51227535  0.95654811        0.15912637  0.51227309  0.95654832
0.15912711  0.51227503  0.95654814        0.15912637  0.51227309  0.95654832
0.15912701  0.51227475  0.95654816        0.15912637  0.51227309  0.95654832
0.15912692  0.51227451  0.95654818        0.15912637  0.51227309  0.95654832
0.15912684  0.51227431  0.95654820        0.15912637  0.51227309  0.95654832
0.15912677  0.51227413  0.95654822        0.15912637  0.51227309  0.95654832
0.15912671  0.51227398  0.95654823        0.15912637  0.51227309  0.95654832
0.15912666  0.51227386  0.95654825        0.15912637  0.51227308  0.95654832
0.15912662  0.51227375  0.95654826        0.15912637  0.51227308  0.95654832
0.15912659  0.51227365  0.95654827        0.15912637  0.51227308  0.95654832
0.15912655  0.51227357  0.95654827        0.15912637  0.51227308  0.95654832
0.15912653  0.51227350  0.95654828        0.15912637  0.51227308  0.95654832
0.15912651  0.51227344  0.95654828        0.15912637  0.51227308  0.95654832
0.15912649  0.51227339  0.95654829        0.15912637  0.51227308  0.95654832
0.15912647  0.51227335  0.95654829        0.15912637  0.51227308  0.95654832
0.15912645  0.51227331  0.95654830        0.15912637  0.51227308  0.95654832
0.15912644  0.51227328  0.95654830        ...
0.15912643  0.51227325  0.95654830
```

Well that's quite enough of that. After appearing to be stable it eventually switches to the genuine stable three cycle.

G Fanoland

THIS FANTASY about a land of love illustrates the fact that the relationships between point and line described in the lecture on Projective Geometry are abstractions and can be mapped onto other things.

If we take a point as a woman and a line as a man and the relationship of a point being on a line as equated with the man and woman having a relationship, then various results emerge. In the first story the seven men and woman map to the points and lines of the Fano plane.

Seven girls and seven boys

ONCE UPON A TIME, seven men and seven women established a friendly commune called Fanoland. The women's names were Ann, Betty, Carole, Daisy, Eileen, Fanny and Greta. The men's names were Alan, Brian, Charles, David, Edgar, Frank and George.

They were trigamous which means that each had three partners. Each man had a relationship with exactly three different women and each woman had a relationship with exactly three different men. However, there was one important rule in their choice of multiple partners. Any two men had just one woman in common among their partners. In other words, given any two men they had five partners between them and thus just one in common. The same rule applied in reverse. Any two women had exactly five partners between them and thus one in common.

Find a possible arrangement of the seven men and seven woman and indicate it by putting a cross (perchance a kiss) in those boxes corresponding to a relationship. The liberty has been taken of inserting one such cross.

	Ann	Betty	Carole	Daisy	Eileen	Fanny	Greta
Alan							
Brian	X						
Charles							
David							
Edgar							
Frank							
George							

Plus six girls and six boys

AFTER A WHILE, the inhabitants of Fanoland became restless for more variety. They recruited six more men (Harry, Ian, John, Keith, Larry and Mark) and six more women (Helen, Irene, Jane, Kate, Lorna and Mary) to join the commune so that there were now thirteen men and thirteen women.

These extra folk enabled the rules to be extended and the commune became tetragamous. Each man now had four woman as partners and each woman had four men as partners. Moreover, the old rule about only sharing one partner was still observed. Thus any two women now had seven partners between them and thus just one in common. And similarly for the men.

There was one sad consequence of the new rules; some of the existing relationships had to be broken.

One day, whilst resting after a particularly jolly frolic celebrating the coming of spring, they were visited by a Princess who noticed a remarkable fact.

They could choose two trios of women such that each trio did not have a man in common. Moreover, if they chose the trios in such a way that they could be paired off (one woman from each trio) so that the three men in common (one man corresponding to each pair of women) themselves did have a woman in common, then they noticed a remarkable fact. They discovered that in these circumstances, the following was also true.

Each trio of women defined a trio of men, one for each pair of women (this was always true of course). If these six men were paired off to form three pairs of men (corresponding to the pairing of the women in a symmetric manner), then the three women shared by these three pairs of men actually always had a man in common.

Prove that this must be so and that it also applies with the roles of men and women reversed.

Explanation

IN THE FIRST STORY the men and woman map onto the Fano plane. Thus we can take Ann as the point 100, Betty as 010, and so on. Similarly, we might take Alan as the line 100, Brian as 010 and so on. Then since the point 100 lies on the line 010 (see the last part of the Finale) it follows that Ann has a relationship with Brian. However, the point 100 does not lie on the line 100 and so Ann does not have a relationship with Alan.

In the second story they map onto the 13-point finite geometry. The rather complex fact noticed by the Princess is simply Desargues' theorem expressed in terms of the relationships.

And on this curious note, this book comes to an end.

Bibliography

The following are referred to in the text. Note that the edition given in some cases may not be the latest but is simply the one in my library. There is of course an enormous amount of material on the web as well.

Abbott, E A, *Flatland*, HarperPerennial (1983)

Baker, H F, *Principles of Geometry, Volume II, Plane Geometry*, 2nd edition, CUP (1930)

Blackett, D W, *Elementary Topology*, Academic Press (1982)

Bollabás, B, *Graph Theory*, Springer-Verlag (1979)

Boys, C V, *Soap Bubbles*, Dover (1959)

Boy's Own Annual, (1917–1918)

Bragg, L, (ed), *The Crystalline State, Volume IV, Crystal Structures of Minerals*, Cornell University Press (1965)

Conway, J H and Guy, R K, *The Book of Numbers*, Copernicus (1996)

Courant, R and Robbins, H, *What is Mathematics?*, 3rd edition, OUP (1946)

Coxeter, H S M, *Introduction to Geometry*, 2nd edition, Wiley (1969)

Coxeter, H S M, *Projective Geometry*, 2nd edition, Springer (1987)

Coxeter, H S M, *Regular Polytopes*, 3rd edition, Dover (1973)

Coxeter, H S M and Greitzer, S L, *Geometry Revisited*, The Mathematical Association of America (1967)

Critchlow, K, *Islamic Patterns*, Thames & Hudson (1976)

Cundy, H M and Rollett, A P, *Mathematical Models*, 3rd edition, Tarquin (1981)

Davenport, H, *The Higher Arithmetic*, 7th edition, CUP (1999)

Devaney, R L, *A First Course in Chaotic Dynamical Systems*, Addison-Wesley (1992)

Devaney, R L, *An Introduction to Chaotic Dynamical Systems*, 2nd edition, Addison-Wesley (1989)

Dirac, P A M, *General Relativity*, Wiley (1975)

Einstein, A, *The Meaning of Relativity*, 6th edition, Metheun (1956)

Fadiman, C, (ed), *Fantasia Mathematica*, Copernicus (1997)

Fadiman, C, (ed), *The Mathematical Magpie*, Copernicus (1997)

Faulkner, T E, *Projective Geometry*, 2nd edition, Oliver and Boyd (1952)

Gamow, G, *Mr Tomkins in Paperback*, CUP (1965)

Gleick, James, *Chaos*, Cardinal (1988)

Grünbaum, B and Shepherd, G C, *Tilings and Patterns*, W H Freeman (1987)

Hofstadter, D R, *Gödel, Escher, Bach*, Basic Books (1979)

Hollingdale, S, *Makers of Mathematics*, Penguin (1994)

Horne, A, *Theosophy and the Fourth Dimension*, The Theosophical Publishing House (1928)

Leonardo of Pisa, *Liber Abbaci* (1202)

de Liaño, I G, *Dali*, 2nd edition, Rizzoli (1989)

Locher, J L, (ed), *Escher, The Complete Graphic Work*, Thames & Hudson (1992)

Luminet, J-P, *Black Holes*, CUP (1992)

Maxwell, E A, *The Methods of Plane Projective Geometry based on the use of General Homogenous Coordinates*, CUP (1946)

Newman, James R, *The World of Mathematics*, Simon and Schuster (1956)

Ogilvy, C Stanley, *Excursions in Geometry*, Dover (1990)

Paccioli, L, *De Divina Proportione*, Venice (1509)

Peitgen, H-O and Richter, P H, *The Beauty of Fractals*, Springer-Verlag (1986)

Peitgen, H-O and Saupe, D (eds), *The Science of Fractal Images*, Springer-Verlag (1988)

Penrose, R, *The Emporer's New Mind*, OUP (1989)

Rindler, W, *Essential Relativity*, 2nd edition, Springer-Verlag (1979)

Rouse Ball, W W, *Mathematical Recreations and Essays*, 11th edition, Macmillan (1939)

Steinhaus, H, *Mathematical Snapshots*, 2nd edition, OUP (1950)

Szpiro, G G, *Kepler's Conjecture*, Wiley (2003)

Thompson, D'Arcy W, *On Growth and Form*, CUP (1992)

Wenninger, M J, *Polyhedron Models*, CUP (1971)

Weyl, H, *Space–Time–Matter*, 4th edition, Dover (1950)

Wilson, R J, *Four Colours Suffice*, Allen Lane (2002)

Wilson, R J, *Stamping Through Mathematics*, Springer-Verlag (2001)

Index

Abbott 27
Appel 117
Archimedean figures 48
   in four dimensions 261
Argand diagram 235

Beecroft 241
bends 240
billiards 217
black holes 225, 226
Boys 137
Brianchon's theorem 103
Buckminsterfullerene 49

cactus 16
Cantor sets 184, 193
catenary 152
chaos 189, 203
   stability 305
Christoffel symbol 222
colouring 117
complex numbers,
   and Newton's method 196
   and geometry 235
compound figures 40, 255
   of five cubes 42, 61, 256
   in four dimensions 282
   stella octangula 40, 252
concentric 157
continued fractions 9, 26
coordinates 104
   homogeneous 104, 244
   in relativity 218
Coxeter 110, 231, 265
crosscap 132
cross-ratio 100
   and bubbles 141
   preserved by inversion 163
cubic equations 315
cube roots 196, 239
cuboctahedron 48, 255
   in 24-cell 280
   and sphere packing 291
curvature 148
   of catenoid 152
   of space 221
cyclotron 216

Dali 24, 35, 86
decagram 30, 59
density 29
   in four dimensions 80
   in three dimensions 46
Desargues configuration 91, 108, 260
   and four bubbles 146
Descartes 241
determinants 245
Devaney 203
diagonal triangle 100, 144
diamonds 298
dihedral angle 38, 56, 265
dipole 115
Dirac 179, 226
Dirichlet region 20
dodecahedron 34
   cubes within 41, 256
duality,
   in four dimensions 72
   in three dimensions 39
   of tilings 31
   in projective geometry 93, 95, 97

Eddington 226, 228
Einstein 206
electrical circuits 8
enantiomorphism 33, 51
   in chemistry 117
   in four dimensions 265
Escher 35, 40, 46, 124
Euler number 40, 47
   in four dimensions 79
   in topology 114, 133

Fano plane 106, 319
faujasite 302
Fatou 199
Feigenbaum 192
Fibonacci numbers 4, 16
finite geometries 105, 320
Fishbourne Roman villa 33
Fitzgerald 206
flatland 27
fluorite 297
fractals 180

Gamow 211

Gödel   61, 179
golden angle   17
golden ratio   3, 14
   the bull and man   247
   in compound of five cubes   42
golden rectangles   44, 256
Gosset   242, 265
graphite   300

Haken   117
Hales   292
harmonic range   100, 141
Hawking   228
Heawood's formula   123, 133
heptagram   29
heptahedron, one-sided   56, 130
honeycombs   64
   in foam   149
   in four dimensions   81, 87
hypercube   71
   cross-sections of   73
   honeycombs   81
   net   84, 267
   projection of   274
   in 24-cell   282

icosahedron   34
   figures within   44
infinity   98, 184
inversion   158

Julia sets   199

Kepler   46, 292
Kerr black holes   228
Klein bottle   127, 134
Koch curve   180

Lami's theorem   139
line at infinity   98, 103
Lorentz   206, 212
Lorentz–Fitzgerald contraction   209, 213

MacMahon's cubes   120
Mandelbrot set   200
Maxwell's equations   179, 218
Meccano   12
Menelaus' theorem   90
Mercury   225
Möbius band   123
Morley's theorem   240
Mr Tompkins   211

Napoleon's theorem   239

nets   84, 267
neutron stars   228
Newton's laws   179, 225
Newton's method   195
non-convex figures   46, 56, 80

octagram   30, 58

Paccioli   24, 40, 44
packing of spheres   287
Pappus configuration   89, 110
particles   208
Pascal's theorem   102
   Hexagrammum Mysticum   110
Pasteur   118
pedal triangles   243
pencil   101
Penrose   60, 228
pentahedron   96
pentagram   13, 29, 59
perovskite   296
perspective   34, 91
Phelan   151
pineapples   21
Platonic figures   35
Poinsot   46
prisms   54, 70
   in four dimensions   265
propositions of incidence   90
Purse of Fortunatus   131
pyramids   82
   in four dimensions   268

quadrilateral   99
   with squares on   231
   and three bubbles   143
quartz   302

rabbits   4, 185
real projective plane   128
rhombic figures   55
   dodecahedron   82, 149, 291
Ricci tensor   222
Riemann   222
Rindler   209
root locus   313

Sarkovski's theorem   192
Schläfli symbol   28, 30, 36
   in four dimensions   69, 87
   for honeycombs   64
Schlegel diagrams   118, 271
Schwarzschild   224
Soddy's hexlet   173, 241

snub figures   51, 265
stability   187, 305
Steiner's porism   157
stella octangula   40, 252
stereo images   59, 251, 274
subdivision of squares   8
sunflowers   19, 24
surface tension   137
symbolic dynamics   195, 204

tetroctahedric   265, 285
tilings   30, 65
   and Archimedean figures   50
   colouring   119
   pentagonal   59
torus   113, 121
trains   207, 210, 213, 229
triacontahedron   55
trisection   240
transformations   51, 52
truncated octahedron   48, 68, 151
truncation   31, 48
   in four dimensions   264

vertex figure   36, 64
   in four dimensions   69

Weaire   151
Weyl   227
Wheeler   226
Whitehead   179
Wilson   24, 134

Zeno's paradox   2
zeolite   301
zonohedra   56

4-simplex   69, 262
   cross-sections of   70
16-cell   71
   cross-sections of   76
   honeycombs   81
   in 24-cell   284
   projection of   276
24-cell   79, 82
   honeycombs   81
   projection of   280
120-cell   79
   projection of   285
600-cell   79
   projection of   285

CPSIA information can be obtained
at www.ICGtesting.com
Printed in the USA
LVHW05*1316280518
578714LV00002B/4/P